Praise for 7

"I have lectured and cons   people understand carrying capacity and its relevance for their communities. Often, I have called upon the teaching tools now shared in *The Climate Change Playbook* in trainings, with staff, in workshops, and in my own presentations. This book is a treasure trove: It is a practical tool kit for any public policy practitioners who want to engage their counterparts and accelerate learning."

—**Mathis Wackernagel**, founder and CEO, Global Footprint Network

"Climate change, like most of our global problems, is a systemic problem—a web of interconnected issues that is difficult to analyze with conventional linear thinking. This book offers a playful, nonlinear, and largely nonverbal, method for learning how to think systemically—in other words, in terms of relationships, patterns, and context. I highly recommend it to anyone who wants to experience systemic thinking firsthand." —**Fritjof Capra**, author of *The Web of Life*; coauthor of *The Systems View of Life*

"Few subjects are more crucial, more discussed, and more poorly understood than climate change. This is a tragedy because there are a few simple, intuitive insights that can be understood by all and could form a consensual foundation that would allow us to focus more clearly on the complex tradeoffs and choices obscured by our misunderstandings. *The Climate Change Playbook* is a great way to understand and, more importantly, help others understand these insights."

—**Peter M. Senge**, senior lecturer, MIT; founding chair, Academy for Systemic Change; author of *The Fifth Discipline*

"Many of us experience the problems of climate change as so overwhelming and beyond our control that we don't know where to start to solve them. This book does the reverse: It makes the issues so palpable that it not only motivates us to do more but also gives us 22 tools we can easily use to mobilize others. If you believe that experience is the best teacher and that we have precious little time to influence changes that have serious long-term consequences for everyone on the planet, this book is an invaluable asset."

—**David Peter Stroh**, author of *Systems Thinking for Social Change*

"One of the major obstacles we face in addressing the climate crisis is the general lack of understanding of the climate system and complex systems in general. *The Climate Change Playbook* provides a novel approach to overcoming this barrier through creative and engaging activities that help move the climate crisis from an abstract threat to a clear and present reality that we can and must act upon today." —**Asher Miller**, executive director, Post Carbon Institute

"In my current work as an environmental scholar and my former work heading the UN-affiliated University for Peace and the International Union for Conservation of Nature, my major goal has always been to help others understand the crucial causes and consequences of environmental issues. These authors are masters of using simple exercises to convey complex issues, and this new book compiles many of their best tools." **—Julia Marton-Lefèvre**, Yale School of Forestry and Environmental Studies; former director general, IUCN

"Thousands of government and corporate officials have participated in training programs that I organize in Japan to convey principles related to environment, climate, food, and energy. I have become a fan of the exercises in the *Playbook*. They are easy to learn and quick to use. They are incredibly effective teaching tools, and they work with participants that do not have English as their first language." **—Riichiro Oda**, president and CEO, Change Agent Inc.

"How can we learn about tough problems like climate change? The research shows that showing people the research doesn't work. To learn, people need to interact, experiment, play. *The Climate Change Playbook* encourages just that through a diverse set of interactive games. Useful with all ages and in groups large and small, these games help us learn critical lessons about difficult topics—and they are a load of fun." **—John Sterman**, professor, MIT Sloan School of Management; author of *Business Dynamics*

"The effort to secure a livable planet for future generations just got a little bit easier thanks to *The Climate Change Playbook*. Whether you are working to educate and empower an audience of students, business leaders, or policy makers, the *Playbook* will help you add interactive learning exercises to your teaching and outreach. With clear and detailed instructions, it is a great resource for anyone working to build sound understanding and a collective will to act on climate change." **—Elizabeth Sawin**, codirector, Climate Interactive

"Using a game to exemplify a point made in a lecture makes all the difference: The audience, large or small, is eager to participate and remembers the message. The beauty of the games in the *Playbook* is their simplicity and flexibility. They can be used with school children, university professors, politicians, and business people, and they lend themselves to debriefing that might consist of a just few sentences or an elaborate discussion. I have become a games enthusiast. The *Playbook* also inspires the creation of variations and even new games to meet specific purposes. We need games to get these vital messages across!" **—Helga Kromp-Kolb**, head, Center for Global Change and Sustainability, University of Natural Resources and Life Sciences, Vienna, Austria

The
Climate Change
Playbook

Also by Dennis Meadows (selected titles)

The Systems Thinking Playbook,
with Linda Booth Sweeney

Limits to Growth: The 30-Year Update,
with Donella H. Meadows
and Jorgen Randers

Also by Linda Booth Sweeney

The Systems Thinking Playbook,
with Dennis Meadows

When a Butterfly Sneezes

*Connected Wisdom:
Living Stories about Living Systems*

Also by Gillian Martin Mehers

Achieving Environmental Objectives,
coedited with Susanna Calvo

The Climate Change Playbook

22 Systems Thinking Games for More Effective Communication about Climate Change

DENNIS MEADOWS,
LINDA BOOTH SWEENEY, AND
GILLIAN MARTIN MEHERS

Chelsea Green Publishing
White River Junction, Vermont

Earlier versions of eighteen games in this book appeared in *The Systems Thinking Playbook*, Linda Booth Sweeney and Dennis Meadows (Chelsea Green Publishing, White River Junction, VT, USA). Copyright © 1995 by Linda Booth Sweeney.

Editor: Joni Praded
Project Manager: Alexander Bullett
Copy Editor: Alice Colwell
Proofreader: Eileen M. Clawson
Designer: Melissa Jacobson

Printed in the United States of America.
First printing May, 2016.
10 9 8 7 6 5 4 3 2 1   16 17 18 19 20

Our Commitment to Green Publishing

Chelsea Green sees publishing as a tool for cultural change and ecological stewardship. We strive to align our book manufacturing practices with our editorial mission and to reduce the impact of our business enterprise in the environment. We print our books and catalogs on chlorine-free recycled paper, using vegetable-based inks whenever possible. This book may cost slightly more because it was printed on paper that contains recycled fiber, and we hope you'll agree that it's worth it. Chelsea Green is a member of the Green Press Initiative (www.greenpressinitiative.org), a nonprofit coalition of publishers, manufacturers, and authors working to protect the world's endangered forests and conserve natural resources. *The Climate Change Playbook* was printed on paper supplied by Thomson-Shore that contains at least 100% postconsumer recycled fiber.

Library of Congress Cataloging-in-Publication Data
Names: Meadows, Dennis L. | Sweeney, Linda Booth, 1963– | Martin-Mehers, Gillian.
Title: The climate change playbook : 22 systems thinking games that teach us how to seek solutions and create change / Dennis Meadows, Linda Booth Sweeney, Gillian Martin Mehers.
Description: White River Junction, Vermont : Chelsea Green Publishing, [2016]
  | Includes bibliographical references and index.
Identifiers: LCCN 2016000955| ISBN 9781603586764 (pbk.) | ISBN 9781603586771 (ebook)
Subjects: LCSH: Climate change mitigation. | Climatic changes. | Educational games.
Classification: LCC QC903 .M435 2016 | DDC 363.738/747—dc23
LC record available at http://lccn.loc.gov/2016000955

Chelsea Green Publishing
85 North Main Street, Suite 120
White River Junction, VT 05001
(802) 295-6300
www.chelseagreen.com

Contents

Preface

Rising emissions of greenhouse gases (GHG) are causing global climate change that has the potential to destroy this planet's capacity to support our species and other life. There is an emerging concern about this threat, yet total annual GHG emissions continue to rise in all countries. They rise in the nations that signed the Kyoto Protocol, and they rise in the nations that did not.

In some years when problems reduce economic growth, the rate of increase may decline slightly. And an individual nation's rate of increase may seem to fall in years when it imports more of its energy-intensive goods from another nation. However, shifting CO_2 emissions from one country to another does nothing to lower the worldwide total. Global emissions still continue to grow.

How is it that emissions keep growing despite rising concern about the climate change they cause? It is possible to identify several reasons for the paradox, most of which lie outside the scope of this book. But one important reason is relevant here: people do not understand the behaviors of the climate system.

And because people are unclear about how the climate system works, they easily make mistakes that are potentially lethal in their consequences. They assume, for example, that when climate change becomes an obvious threat to human society, there will still be time to make changes that will avoid disaster. Yet at that point it will be too late. They assume we can use our current paradigms and policy tools to find solutions. Yet the approaches that caused damage in the first place will cause even more damage in the future. They assume that the relevant consequences of an action will appear near in time and space to the action itself. Yet actions are often destructive elsewhere and later. Scientists' predictions have not altered people's assumptions, since warnings and appeals are weak foundations for learning. As the old proverb goes

> *When I hear, I forget.*
> *When I see, I remember.*
> *When I do, I understand.*

Simple interactive activities—we call them "games" or "strategic exercises"—have the potential to help participants understand by doing. In *The Climate Change Playbook* we present twenty-two games useful to those trying to help others understand and deal more effectively with climate change. These games facilitate the processes of communication and teaching.

This book is an adaptation of *The Systems Thinking Playbook*, first published in 1995. We have modified eighteen of the original games and added four new ones. The game descriptions are now specifically focused on helping people perceive climate change dynamics and anticipate its consequences.

How to Use This Book

This book was created to help experts, advocates, and educators be more effective when talking with groups about climate change. If they are used well, the games can make workshops, speeches, and conversations on this complex topic more productive. Each of the twenty-two games has practical guidelines for facilitators. The game descriptions include the following:

Quotes: Relevant quotes related to climate change and the game dynamics.

Climate Link: Specific information related to climate change, to be used for framing the game and considering the most effective context.

About the Game: Helpful information about the history of the game and strategies for making it most effective as a tool for teaching.

To Run This Exercise: Requirements for using the game, including the following information: number of people, time, space, equipment, and setup requirements. Things to consider, such as safety or physical limitations, are also noted where appropriate.

Instructions and Script: Detailed instructions for engaging in each exercise, including specific script suggestions for the facilitator.

Debrief: Discussion points to maximize the learning potential and takeaway value of each game.

Selecting the Game You Need

Sessions related to climate change involve different numbers of participants. Thus we divide the games in this book into three functional categories, according to the number of people involved directly in the action of the exercise:

1. **Mass games.** These games can be played by a very large number of participants, up to many hundreds. They normally do not require participants to interact with each other. Instead, each person in the audience listens to instructions and questions from the facilitator. These exercises can be carried out in large conference settings while participants remain in their seats. A single facilitator can lead thousands of people through a mass game at the same time.
2. **Demonstration games.** These games involve a small group of people, usually less than ten. They interact with each other while being observed by a very much larger group. The audience can be of any size, so long as everyone can see the action of the game.
3. **Participation games.** These games may be used for larger groups, up to thirty. Because it is difficult to learn vicariously from the action in these exercises, it is more effective to involve everyone in the group directly in the play.

System Behaviors and Misconceptions

Confusion about climate change results from a variety of misconceptions. Thus we index the games in this book according to six features of the climate system that cause special difficulties for those who wish to understand and address it.

1. **Habitual behavior.** Climate change results from actions that have become deeply embedded in the habits of global society. Some actions that used to be beneficial now threaten the

survival of our species. Efforts to reduce climate change will succeed only if these habits are changed.

2. **Inappropriate frames.** Problems caused by climate change, such as a melting icecap in the Arctic or floods in Pakistan, occur in places very far away from the actions that produced them. Efforts to reduce climate change require enlarged perceptions of time, place, and sense of responsibility.

3. **Uncertainty.** Climate change is caused by a complex set of interactions that are not entirely understood or accurately measured. Efforts to reduce climate change require groups to discuss and reach consensus about novel problems that they do not thoroughly understand.

4. **Autonomous behavior.** Climate change is brought about by an intricate system structure not fully under human control. It contains processes that can escalate on their own. Actions to reduce climate change must reflect respect for the natural systems at work, independent of social control.

5. **Long delays.** Climate change involves processes that contain very long delays. The full consequences of past emissions have still not been fully experienced. Even after appropriate policies are implemented, problems will persist for many decades. Efforts must surpass resistance in the short term and advance goals for the future.

6. **Magnification.** A seemingly small change, for example, raising CO_2 concentrations in the atmosphere by a few parts per million, can cause major problems, such as the permanent loss of a species. Efforts to reduce climate change must consider signals that do not initially seem important or significant.

Using the Games Matrix

To guide you in selecting a game best suited to your circumstances and goals, we have developed a games matrix. Each exercise is classified primarily as a mass game, a demonstration game, or a participation game, and we indicate which features of the climate system it may be used to illustrate. These classifications are not absolute. As you master the games, you will find you can use them in many ways.

The Games Matrix

Name	Function			Misconception
	Mass	Demonstration	Participation	
Arms Crossed	X			1
Avalanche		X		3, 4
Balancing Tubes			X	5, 6
The Bathtub Game		X		2, 5
Biodiversity Game	X			2, 4, 6
Circles in the Air	X			1, 2
Frames	X			2
Group Juggle			X	3, 4, 6
Hands Down	X			2
Harvest			X	1, 2, 3, 4, 5
Hit the Target		X		5
Living Loops		X		4, 5
Paper Fold	X			4, 6
Paper Tear	X			2, 3
Pens	X			2, 3
Space for Living			X	1, 2, 3
Squaring the Circle			X	2, 3
Thumb Wrestling	X			1, 2
Triangles			X	4, 6
Warped Juggle			X	3, 6
Web of Life		X		2, 4, 5
1-2-3-Go!	X			1, 3, 4, 6

Things to Consider before Getting Started

Close physical proximity is involved in many of the games. Being close to strangers, especially of a different gender, may be disturbing to some people. Often participation in a game involves holding hands or standing together. If one or two people seem to feel uncomfortable with an exercise you intend to use, enlist them to help in some other way—by checking for compliance with the rules or by watching and reporting their observations, for instance. If three or more participants will be uncomfortable with a specific exercise, change it or do not use it. Sometimes asking each participant to hold an end of a napkin can reduce physical contact between partners. If male-female contact is an issue, consider dividing the participants into a male subgroup and a female subgroup.

Many of these games are run with the participants seated or as a demonstration with most participants playing the role of observer. Several of them, however, require the active participants to stand, move, or walk among the other participants. Watch to ensure that no game places anyone under physical stress at any time or poses the potential for someone to lose his or her balance and fall. These exercises have been run hundreds of times with no problems, but it is always wise to respect your participants and be cautious. If one or two participants have mobility limitations, enlist them to help in some other way, adapt the game, or use a different game.

During discussions and the debriefing sessions, it is best not to call on a specific person. Putting someone on the spot might be embarrassing. Allow participants to choose to share their thoughts or not.

Debriefing Guidelines

Debriefing may take a few minutes, or it may become an extended discussion. But there absolutely must be some conversation about the game and its relation to the more general purposes of your session. Either the facilitator or the participants should summarize the main experiences and insights immediately after each game.

We've outlined seven steps to provide some practical guidance about the debriefing process. After each game session, your debriefing may follow all seven steps, or you may choose to skip some of these steps or to condense several of them into one phase of your conversation with the participants:

1. Describe the events and issues that occurred during the game.
2. Determine the extent to which those events and issues also occur in the real system.
3. Decide what factors in the game were responsible for those events and issues.
4. Determine the extent to which those factors are also present in the real system.
5. Identify changes in the game that would avoid or solve the most serious issues.
6. Indicate how corresponding changes could be made in the real system.
7. Gain commitment to actions that will achieve the necessary changes in the real system.

Ensure that the players feel a sense of responsibility for the behaviors that occur in their game. This is critically important. If they attribute their failure to some exogenous influence, random variable, or a mistake by the facilitator, they will have no incentive to examine the game result and learn from it. Through your careful facilitation, you can help them to become students of their own behavior, learning from their own mistakes and successes. But you must do this in a way that avoids embarrassment. Do not let participants imagine that you consider their behavior to have come from stupidity, ignorance, or ill will. You can employ this kind of language, for example: *"When a group of intelligent people like you, who honestly want to do well, behave in this way, there must be an underlying reason."*

Guiding Ideas

"Systems thinking" is a broad term used to represent methods that focus on collections of elements and their interconnections—rather than on individual parts. It provides a context for defining and solving complex problems, therefore fostering more effective learning and design. At its best, the practice of systems thinking helps one to stop operating from crisis to crisis and to think in a less fragmented, more integrated way.

The games in this book highlight many of the concepts and habits of mind associated with systems thinking, providing novel insights into the dynamics of climate change.

In the twenty-first century, society faces an important educational challenge: learning how to help people become more effective at understanding and coping with increasingly complex systems. More and more, practitioners and academics adhere to a simple premise when designing learning experiences: engage the mind and the body. In their powerful book, *An Unused Intelligence*,[1] Andy Bryner and Dawna Markova warn that the Western culture of education leaves the problem-solving potential of human bodies virtually untapped. We wholeheartedly concur, and we add that the systems-thinking and systems-sensing potentials of our bodies have been untapped as well. What you experience using these exercises will depend on the skillful integration of key concepts, theory, techniques, and experiential exercises; your familiarity with systems-thinking concepts; and the insight and energy of the facilitator. Experientially, you will raise awareness of the habits of mind found in a systems thinker. And we invite you to have some serious fun while you're at it.

We have tried to set up *The Climate Change Playbook* so that anyone—managers, CEOs, teachers, and professors—can read it and use it and find something meaningful. You don't need to be an organizational development professional or trainer to apply these games. In fact, we envision that with a bit of preparation any team will be able to open the *Playbook* and work through the exercises.

The Ways of Systems Thinkers

Our experience studying and teaching systems thinking has led to the definition of systems thinkers as those who:

1. See the whole picture
2. Change perspectives to see new leverage points in complex systems
3. Look for interdependencies
4. Consider how mental models (a person's beliefs, ideas, assumptions about how the world works) create our futures
5. Pay attention to and give voice to the long term
6. "Go wide" (use peripheral vision) to see complex cause-and-effect relationships
7. Find where unanticipated consequences emerge
8. Focus on structure (the interrelationships within a system), not blame
9. Hold the tension of paradox and controversy without trying to resolve it quickly
10. Make systems visible through causal maps (which show how various actions influence other actions and outcomes) and computer models
11. Seek out stocks (accumulations of materials and information within a system) and the time delays and inertia they can create
12. Watch for win/lose mind-sets, knowing they usually make matters worse in situations of high interdependence
13. See themselves as part of, not outside of, the system

These exercises are meant to promote a greater awareness of these ways of thinking and seeing when learning about climate change. They are best used within an interrelated and reinforcing design that covers theory, concepts, and models and includes a relevant and detailed debriefing of the participants' experience.

22
Games

1.

Arms Crossed

When conditions change, habits must change.

Quotes

First we form habits, then they form us. Conquer your bad habits or they will conquer you.

—Rob Gilbert, sports psychologist

The problems we have created in the world today will not be solved by the level of thinking that created them.

—Albert Einstein, theoretical physicist

Bad habits are like a comfortable bed: easy to get into but hard to get out of.

—Proverb

Climate Link

The accumulation of greenhouse gases in the atmosphere continues, because human society has developed habits related to economic and demographic growth. These habits will produce those gases at an ever-increasing rate. In order to reverse climate change, we must change these habits. No matter how dangerous or dysfunctional the current habits are, there will be important people and organizations that fight against any attempt to modify them. Successful efforts to combat climate change will require society to develop many new habits influencing its consumption, transportation, politics, energy use patterns of urban design, and so on. This exercise alerts us to some of the challenges we face in any effort to change our habits.

About This Game

Arms Crossed is useful because it is quick, it requires no special equipment, and it works well with audiences for whom English is not their first language.[2]

To Run This Exercise

Number of People

This is a mass game. It can be used with audiences ranging from a few people to a thousand or more. Expect that everyone in your audience will participate.

Time

The exercise takes several minutes. The debriefing discussion can occupy a few minutes or much longer. You can choose the option that best serves your teaching goals.

Space

This exercise is suitable for a large seated audience. It is only necessary that every participant can see and hear you.

Equipment

None.

Setup

None.

Instructions and Script

Step 1: Say to the group, *"Now I am going to lead you through a brief activity. I need for everyone to participate. So anyone who is holding something—a pencil or pad of paper, for example—should put it down."* Look around the group and ensure that everyone is ready to follow your instructions. If you see people still holding things, again ask that they set the things aside.

"Everyone fold your arms." As you say this, fold your own arms. *"Now look down and make a mental note about which wrist is on top, and remember whether it is the left or the right."*

"Now drop your arms." Drop your own arms to your sides in order to illustrate what you want them to do. Pause for a moment.

"Now cross your arms again. Look down and remember which wrist is on top."

Pause long enough for them to accomplish both of your instructions. *"Now drop your arms."*

Step 2: Tell the group, *"Now we will conduct a little survey. Everyone who had the same wrist up both times, raise your hand."* You should raise your own hand to illustrate what you want while saying to the group, *"I had the same wrist up both times."* (Of course, you need to make sure that you did actually fold your arms the same way both times.) Normally, all but a few people will raise their hands. Look around the group. Remark, *"Almost everyone had the same wrist up both times. But that is desirable. Folding your arms is what you do when you need to focus your attention on something that does not require your arms. Once you find an action that gets your arms comfortably out of the way, you use it whenever it is required. It would be a big* *waste of time if you had to start from the beginning to find something to do with your arms every time you wanted them out of the way."*

Step 3: Point out, *"Since almost everyone folds his or her arms in the same way, time after time, there must be an optimum way to do it. Let us see what it is."*

Next, see how many people cross their arms the way you do. In these instructions we will assume you, the facilitator, had your left wrist on top each time.

"Everyone who had the left wrist up both times, raise your hand." You raise your hand and remark, *"I did."* Then drop your hand.

"Everyone who had their right wrist up both times, raise your hand."

Look around the group. Ignore the scattered few who folded their arms differently each time and thus did not raise their hand after either of your preceding questions. Normally, it will be about 50 percent one way and 50 percent the other.

"About half of you have the habit of doing it one way, and half have the habit of doing it the other way. There is no optimum. You could do it either way. But once you find a habit that works, you just keep using it without question. It may even not occur to you that there is a different way to do it and that many people around you use another way."

Step 4: Observe, *"We adopt habits because they are effective. As long as they are effective, we can use them automatically. We do not have to think about them. But sometimes conditions change. Then a habit that was effective is no longer useful. Then the habit must be changed. I am going to help you practice changing your habit.*

"Everyone cross their arms the other way." Do it yourself, making some exaggerated effort to show that it requires thought and perhaps an initial mistake.

Wait for ten seconds. Typically, there will be some nervous laughter from the group until everyone has managed to fold their arms one way or another.

"Congratulations! You did it! But notice three things that are always true when we change our habits. First, it is possible. You did

all manage to change your habitual way of crossing your arms." Pause and let them consider your statement. *"Second, it was not easy. It required thought and probably some initial mistakes."* Pause and allow them to reflect. *"Third, it is uncomfortable at first. You all felt a little strange doing it differently than you normally do.*

"For more than 250 years, humanity was typically made better off through actions that promoted economic growth and increased our consumption of energy and materials. We developed an extremely effective set of habits for achieving increased energy use, raising food production, harvesting the forests, and so on. Now circumstances have changed. Actions that promote economic growth and raise our consumption of energy and materials even further will typically make us worse off over the long term.

"To sustain human welfare, we need to reduce our impact on the climate. We need to reduce our use of activities that put more greenhouse gases into the atmosphere. We have to change our habits. Attempts to do that will show us once again: First, it is possible. Second, it requires careful thought and entails inevitably some initial mistakes. Third, some people and some institutions will resist changes that they think will make them worse off over the short term. There is no solution for climate change that is going to make everyone happy."

Debrief

- What societal habits seem most influential in producing rising greenhouse gas emissions?
- Are those habits necessary, or would it be possible to behave in a different way?
- If you accept the three lessons from Arms Crossed, how would you interpret and apply them to have the best chance of changing society's habits?

This exercise can be used with no formal debriefing. The closing words above may be sufficient.

2.

Avalanche

Understand the implicit rules. They can produce different results than desired or expected.

Quotes

If you want to understand the deepest malfunctions of systems, pay attention to the rules and to who has power over them.

—Donella Meadows, environmental leader

Learn the rules so you know how to break them properly.

—Dalai Lama, spiritual leader

Laws control the lesser man. Right conduct controls the greater one.

——Mark Twain, humorist

Climate Link

Governments say they will reduce their nations' greenhouse gas emissions, but the emissions keep going up. People claim they are concerned about the long term, but their actions promote short-term gains even when they will produce long-term costs. Politicians campaign on the promise to do something about climate change, but when they are in power they may take actions that make it worse.

The rules of the system are producing outcomes different from what many people claim they want. The rules are embedded in laws, administrative custom, and cultural norms. As long as those rules are in force, they will accelerate climate change. To reduce greenhouse gas emissions, we will need to change the rules. This exercise shows that when a set of rules produces a problem, merely trying harder will not solve the problem as long as the effort is enacted within the same rules that caused the problem in the first place.

About This Game

This exercise requires some practice in order for it to produce the desired results. When you know how to conduct it, you will have a powerful object lesson for your participants. Most negotiations are based on the implicit assumption that if everyone agrees on a goal and works hard to achieve it, success will follow. The Kyoto Protocol is based on this assumption, for example. But success does not always follow. Often the implicit rules of the system produce a result very different from what people want and expect. Using this exercise, the results may be very graphic and *very* surprising. If your colleagues think that they only need to "educate" people in order to start dealing with climate change, this exercise may jolt them out of that complacent frame of mind.

To Run This Exercise

Number of People

This is a demonstration exercise. Conduct one session in front of a large audience. Seven people are required to carry out the game. You may want to ask for volunteers. If none step forward after about five seconds, simply point to seven people and ask them to join you. Pick these people from the front of the audience in order to minimize the time lost while people move to the game area (the stage or other clear space).

Time

Ten minutes to conduct the exercise and ten to thirty minutes to discuss.

Space

You need enough room that you can stand in the middle of a tight circle of seven people in front of and in plain view of all the other members of the audience.

Equipment

You will need a hoop. A plastic hula hoop of 30–35 inches (75–90 cm) in diameter works fine. If you need to carry the hoop in your luggage, it is best to buy one that can be assembled from pieces, such as the Expand-O-Hoops kit, available online from school specialty stores.

Setup

Assemble the hoop, if need be, and place it near the podium so that it is easily available to you when you are ready for the exercise.

Instructions and Script

Because Avalanche is a demonstration game, you may be introducing it within the context of a presentation or broader program. At the appropriate time, explain to your audience that you are going to conduct a brief exercise that illustrates an important fact about climate change.

Step 1: Hold up the hoop. *"I ask you to imagine that this hoop represents the level of CO_2 in the atmosphere. Our goal will be to lower it as quickly as possible. Starting here* [hold the hoop at waist height] *a team of us will work to lower it, taking it down to the level of the ground. I need a team of seven people to accomplish this."*

Ask for volunteers, or simply point to seven people in the first row or two of the audience. Make sure your group does not include anyone who would have physical difficulty bending over or kneeling down on the floor. All the group members should be able to communicate in one shared language. Ask the group to stand and join you in front.

Step 2: *"In just a moment I am going to ask you to work as a team in order to reduce these emissions; that is, to lower the hoop to the level of the ground.*

"There are two rules that you will all need to follow. Please pay attention as I describe them. They are extremely important." Inform them that you will be watching carefully to make sure that everyone follows the rules. Explain that if anyone violates a rule, you intend to point them out to the rest of the group and ask them all to start the exercise again from the beginning without stopping the clock.

"First, hang your right arm down, so that the elbow is near your waist. Extend your right hand out in front of you, palm down, make a fist and extend your index finger." Do this yourself to illustrate what you want.

"Please stand around me in a small circle. In a moment I will lower the hoop down until all of you are supporting it, each of you only with the top of your index finger touching the hoop."

Again, emphasize: *"There are two rules. The first rule is that each person can touch the hoop only with the top of his or her finger. The second rule is that no one must ever, ever lose contact with the hoop, not even for a fraction of a second! This is a team effort. And if any of you loses contact with the hoop, it means you are not doing your fair share of the work. If I see that any one of you loses contact, I will immediately stop the exercise and start again from the beginning without stopping the clock.*

"When you are ready, I will let loose of the hoop and say, 'Go!'
Then someone in the audience will time you to see how long it takes
you to complete the task of lowering CO_2 emissions, the hoop, all the
way to the floor. I will stay inside your circle, thus inside the hoop.
You can just lower it as quickly as you can with me standing inside
the circle."

Typically, one member of the group will ask you whether it is
permitted for them to talk to each other. The answer is, *"Of course!"*

The seven group members should be more or less equally spaced
around you. Lower the hoop down to the level of your waist and
ensure that everyone is touching it with the top of his or her finger.
You should not be grasping the hoop; you can push down on it semi-
firmly from the top with one finger of each hand.

Step 3: *"Before we start, I will ask the audience: How long do you*
think it will take for this fabulous team to accomplish their goal?"
Give the audience a chance to ponder your question. This is very
important, since their ability to estimate a time for the task establishes
that it is plausible. Solicit answers from one or two members without
offering any comment. Then ask for someone in the audience to use a
watch with a second hand to time the effort. Now look around at the
members of the seven-person team, ensuring that everyone is touching
the hoop.

Step 4: *"Remember, each of you, it is **extremely** important not to lose*
contact with the hoop. Are you ready? Go!" At the moment you say,
"Go!" raise your fingers up from the top of the hoop, letting go of it
completely, and center yourself inside the hoop to make sure that the
hoop does not touch you as the group strives to lower it.

As soon as you quit pressing down on the hoop, it will probably
start to rise up. As soon as it rises to the level of your head, or higher
if it is moving fast, grab it. Sometimes the hoop remains more or less
level as the participants struggle to understand what to do and how
to coordinate their efforts. Watch the hoop carefully. Almost always
you will see someone lower his or her finger and momentarily lose

contact with the hoop. If that happens, point it out. In a humorous way chastise the person who violated the rule and make a big show of bringing the hoop back to the level of your belt and starting again. Eventually, the hoop will start to move up, accelerating as people try to maintain contact with it by pressing up. At some point you should have the opportunity to grab onto the hoop as it is moving upward around the level of your head. Then you can say, *"Okay. We need to stop now. Thank you all for a terrific effort. Please sit down."*

The group's attempt to lower the hoop will fail almost every time. In very rare cases, however, the group does succeed in lowering the hoop to the ground but only by moving very, very slowly. Taking a great deal of time is also a failure, though of a different sort. One way or the other, you have a "failed" effort for use in the debrief.

Wait until participants are seated and then begin the debriefing.

Debrief

"What happened?" Give your audience a moment to think about your question and formulate an answer. *"The goal of the group was to take the hoop down to the ground. What actually happened to the hoop?"* Let someone give you the obvious answer: *"The hoop went up instead of down."*

Now you need to make sure they are not embarrassed by their failure. *"When a group of smart, dedicated, well-intentioned people like our team here fails to do something that they all want to do, there must be a systemic reason. Why did that happen here?"* Give your audience time to reflect and venture opinions.

Here is the key lesson: *"It happened because the rules produced a different result from the one we wanted. Those who created the rules and those who followed them may not have intended this to happen, but it did. People often develop rules that produce results different from what they desire."*

Offer the following explanation: *"So what were the two rules that frustrated us here? First, team members could touch the hoop only with the top of their finger, and second, no one was permitted to lose contact with the hoop. As long as those rules are in effect, success is practically impossible. You could send everyone to an athletic club to exercise and strengthen their fingers. You could hold team-building sessions that encourage everyone to work together. You could have a daylong meeting for people to share their views. You could conduct workshops to improve the team members' communication skills. You could do many other things, but the results would be the same as long as the rules did not change.*

"Though everyone understood the rules, and the rules seemed reasonable, they prevented success. Why? The reasons are easy to describe. In order to maintain contact with the hoop, each member needs to press his finger up against the bottom of the hoop. That tends to raise the hoop. When it rises, other members of the team need to raise their fingers up to maintain contact, thereby raising the hoop even farther. It is a vicious cycle, and the hoop goes up instead of down.

"It would be quite easy to change the rules to get the desired result. You could, for example, let people pinch the hoop instead of resting it on the tops of their fingers. Or you could say that it is okay for someone to lose contact with the hoop. But you need to make some rule change in order to achieve the stated goal."

- What are some of the rules that seem to govern society's response to rising levels of CO_2 and other greenhouse gases in the atmosphere? How are these rules communicated?

In Avalanche the game is played according to explicit rules. In real life those "rules" are more vague—habits or cultural norms. In our society one "rule," or belief, is that if you can't see it, it won't hurt you. The levels of CO_2 and other greenhouse gases that cause damage to the planetary climate are a few hundred parts per million. The gases are invisible, and at a few hundred parts per million they seem too small for concern. The inaction is reinforced by mass media and decision makers who have a vested interest in the status quo and criticize any observation that might challenge it. Another "rule," built into democratic decision making, is mainly to worry about issues that will cause problems between now and the next election. Rising greenhouse gases do not seem to do that. This fact is also communicated by the media.

- Will we be able to avoid climate change while those rules are observed?
- How could the rules be changed to give a better chance for reducing greenhouse gas emissions?

3.

Balancing Tubes

You can't achieve long-term goals with short-term perspective.

Quotes

The appropriate time horizon for socioeconomic scenarios depends on the use to which they are put. Climate modelers often use scenarios that look forward 100 years or more. Socioeconomic scenarios with similar time horizons may be needed to drive models of climate change, climate impacts, and land-use change. However, policymakers also may wish to use socioeconomic scenarios as decision tools in framing current policies for climate change adaptation. In this context, time horizons on the order of 20 years may be more appropriate, reflecting the immediate needs of decision makers.

—Intergovernmental Panel on Climate Change

Advocates can try to frame global warming in a way that makes it seem like the kind of "here and now" crisis we are familiar with, or they can do the much harder work of reframing value systems so that we do something rare for our species: act now to limit risks facing our children and their children.

—Andrew Revkin, *New York Times* science journalist

If you are planning for a year, sow rice; if you are planning for a decade, plant trees; if you are planning for a lifetime, educate people.

—Chinese proverb

Climate Link

In policy challenges involving an immense number of causes, effects, and unintended impacts—such as climate change—our typical approach to decision making fails in part because we never experience the impact of our decisions. When considering actions or research related to climate change, it is important for groups to agree on the time horizon of a specific study or action. For instance, over what time frame do we expect high-volume emitters of carbon dioxide to change behaviors, particularly when there are apparently no immediate consequences to their actions? Agreeing on time horizons can help groups to avoid the miscommunication, misunderstanding, and conflict that arise when group members have implicitly adopted different time frames for an issue.

About This Game

Using handmade tubes, this exercise offers a physical experience from which participants can increase their awareness and understanding of appropriate time horizons, particularly as they relate to climate change. This exercise does not convey all aspects of the concept, but it does make this important point: when you are trying to understand and control a dynamic system, there will be an appropriate time horizon within which your observations can lead to insight and effective management. Focusing on changes that occur in the system over shorter or longer periods than the appropriate time horizon will not give you control.

To Run This Exercise

Number of People

The number of participants is limited only by the quantity of tubes you can prepare.

Time

Five minutes for the exercise, ten minutes or longer for debriefing.

Space

3 to 4 feet (1 m) between each person. Have the group stand in a circle so each person can see the performance of the others.

Equipment

One paper tube, approximately 1 inch (2.5 cm) in diameter and 3 feet (1 m) long, for each participant. Alternatives are sticks or cardboard tubes of similar dimensions.

Setup

Prepare enough tubes ahead of time. Using a sheet of newspaper or newsprint, begin at one corner and roll diagonally around a broom handle. Slide the paper off and tape it. You might place one tube at each seat prior to the arrival of participants. Or store the tubes in a paper grocery bag and quickly pass them around just before beginning the exercise.

Instructions and Script

Step 1: Tell participants, *"Your goal is to balance this tube on your fingertips."* With your palm facing up, demonstrate balancing the tube vertically on your fingers. *"First balance the tube while focusing your*

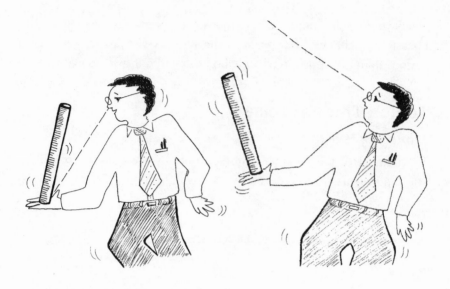

eyes on a spot just 1 inch (2.5 cm) above the point where the tube meets your fingers." Pause to give the group time to try this.

Step 2: "*Now balance the tube while focusing your eyes on a point at the top of the tube.*" Pause while the group tries this.

Step 3: "*Finally, try to balance the tube while focusing your eyes on the ceiling.*" Wait for your group to try. Participants will find it difficult or impossible to balance the tube of newspaper on their fingers when looking at a spot that is either too close to their fingers or too far away.

Debrief

Here are a few questions to begin with:

- Which of the three methods worked best?
- Why do you think it was easiest to balance the tube when focusing on the top of the tube?
- What was changing when you shifted your perspective?

The main factor that changes when you shift from one focal point to another is the length of time between when the tube starts to fall off balance and your eye detects the movement and then provides the information required for your hand to adjust. This is because the tube must move a certain distance before your eye can detect that it has changed position. This is sometimes referred to in psychological experiments as the just noticeable difference, or JND.

When you focus on the bottom of the tube, the top must move a great distance to provide the JND stimulus you need to counteract the fall. Typically, you will be too late, and the tube will fall. When you focus on the top of the tube, the top needs to move only a little to give you the JND movement you need to cope. So your response is relatively quick and usually effective in maintaining balance. Of course, when you focus on the ceiling, the tube can fall almost completely off your finger before you notice its movement, and you have practically no control at all.

Here is the key lesson to take from Balancing Tubes: If you want to control something, the time horizon you choose must be compatible with the dynamics of the systems. If your time perspective is too short or too long, you will not be able to control satisfactorily the behavior of the system.

As a group, consider the climate change system.

Ask the group: "*If we take action now to reduce CO_2 emissions, where are we most likely to see results?*" Pause for an answer. "*The answer is that we will see a reduction of CO_2 being put into the atmosphere and a reduction of CO_2 to the oceans and other sinks. Now consider the time horizon. If we do something today to reduce CO_2 emissions, it will likely take many decades to see changes in wind velocity, precipitation, ocean level, and the resulting storm, flood, and drought damage that changes in these factors will cause. We tend to look where we can make changes and avoid the politically difficult or uncertain areas.*"

Ask, "*What long-term climate change policies are currently being made on the basis of short-time horizons and short-term considerations?*"

You can also make the point that wait-and-see attitudes are the equivalent of focusing your eyes on the ceiling. Here is one way you may opt to discuss wait-and-see attitudes:

"Some situations, such as boiling water for tea, have short time delays and very clear time horizons. In boiling water for tea, one can imagine the time horizon to be approximately five minutes. The delay between action (filling the tea kettle and bringing the water up to a boil) and results (a cup of tea) is very small. Few complex public policy challenges have such short delays and time horizons. Yet surveys show that among people who believe that climate change poses serious risks, many also believe it is safe to delay reductions in greenhouse gas emissions to a level sufficient to stabilize atmospheric greenhouse gas concentrations until there is greater evidence that climate change is harmful. Indeed, many policy makers around the world argue that it is prudent to wait and see whether climate change will cause substantial economic harm before undertaking policies to substantially reduce emissions.[3]

"Such wait-and-see approaches are dangerous because they (1) underestimate the substantial time delays in the climate's response to the consequences of human emissions and (2) presume that climate change can be reversed quickly when harm becomes evident through, for example, changes in ice cover, sea level, weather patterns, and agricultural productivity or changes in the distribution of species, extinction rates, and the incidence of diseases."

4.

The Bathtub Game

A level will decline only if outflows are greater than inflows.

Quotes

Widespread underestimation of climate inertia arises from a more fundamental limitation of people's mental models: weak intuitive understanding of stocks and flows—the concept of accumulation in general, including principles of mass and energy balance.

—John Sterman, director, MIT System Dynamics Group, and Linda Booth Sweeney, systems educator

Confusion about the interrelation of greenhouse gas (GHG) emissions and accumulated GHG amounts is one of the most important sources of mistakes affecting social policy responses to climate change.

—Dennis Meadows,
emeritus professor of systems policy

Climate Link

People with a wait-and-see attitude toward climate change assume that climate change impacts can be addressed or reversed quickly when harm becomes evident. They would defer any action until there is clearly a problem. This view grossly misjudges the climate's response to human GHG emissions.

To understand the mistake, consider the metaphor of a huge bathtub. Whenever water is flowing into the tub faster than it is flowing out, the amount of water in the tub increases. Note that it will *continue* to increase even if the faucet is turned down a little, reducing the inflow, if that diminished inflow still exceeds the outflow. Similarly, the accumulation, or stock, of greenhouse gases in the atmosphere will continue to rise so long as combined emissions exceed combined withdrawals, even if the emissions are being lowered.

Presently, the combined emissions of GHG are at least two times greater than their combined withdrawal through all biological and geochemical mechanisms; for example, through biomass uptake like forest carbon sequestration, ocean absorption, and weathering of rocks. Thus, inflows must be reduced substantially before the accumulation will start to go down. Minor reductions will only slow the rate of growth of the problem. It will require a long time, decades, to accomplish the changes necessary for a large reduction in GHG emissions. We need to start that process long before its need is apparent from the actual behavior of the climate.

About This Game[4]

In this activity, participants have the opportunity to experience physically the rise and fall of CO_2 in the atmosphere. Using a large, marked-off area on the floor, participants enact the inflow and outflow of a stock (like water in a bathtub) and predict the changing level of stock over four rounds of play. They then look at how the same stock/flow structure may help people to make predictions related to CO_2 emission policies.

Depending on the group knowledge of climate change dynamics, you may want to discuss the source of CO_2 emissions and provide current information about the accumulation of CO_2 in the atmosphere and sources of CO_2 removal, or outflows.

To Run This Exercise

Number of People

This is a demonstration game using seventeen participants.

Time

Fifteen to thirty minutes.

Space

Large enough for sixteen people to stand and walk in and out of an 8-by-8-foot (2.5-by-2.5 m) square marked off with masking tape on the floor.

Equipment

White board, chalkboard, flip chart, or easel pad;
Colored markers;
Making tape or other colored tape to create a space for the activity on the floor (alternatively, use rope or string).

Setup

Mark off a large square on the floor with tape. This is your bathtub, or stock.

Create four graphs on a flip chart or other large writing surface. Title the graphs Trial 1, Trial 2, Trial 3, and Trial 4. On each graph, label the vertical axis "People in the stock" and number it, starting at the bottom, from 0 to 20. Label each horizontal axis "Cycle" and number it, from left to right, 0 to 5.

Instructions and Script

Step 1: Ask six people to stand in the square (the bathtub). The rest of the people should stand in a group close to but outside of the bathtub, or stock.

Tell the audience, *"The people in the bathtub represent the accumulation (or stock) of CO_2 in the atmosphere. The people **entering**

the bathtub represent the rate of CO_2 emissions. The people **exiting** the bathtub represent the drain-out of CO_2, which could be, for example, through absorption by biomass."

Say, "We'll play this game with four different trials. In the first trial, two players will enter the bathtub and no players will leave each cycle. We will repeat this action—two players in and no players out—for a total of five cycles."

Step 2: Ask for a volunteer: "I need one volunteer to graph the results." Tell this person to keep track of the level of stock—the total number of people in the bathtub—for each trial. Turn the flip chart away from the group.

"Before this first trial begins, each of you please make a prediction. Write this down on a piece of paper and don't discuss your prediction with anyone else. Do you think the stock will increase, decrease, or stay the same if more players are going in than going out?"

Trial 1: Two players in, no players out for each of the five cycles.

Step 3: Say, "Okay, let's try it!" Instruct two players to walk into the bathtub and have no players leave. Continue that proportion—two in, none out—for a total of five cycles.

Pause and ask: "What is changing here?" Ask a few people to share their predictions. Ask the person drawing the graph to reveal his or her graph, which should show sixteen people in the stock at the end of cycle 5.

Repeat this same process for three more trials. Always start with your stock of six people in the bathtub but change the proportions:

Trial 2: For two cycles, two players in with no players out. For the last three cycles, one player in, no players out.

Trial 3: Two players in and two players out for each of the five cycles.

Trial 4: Two players in for one cycle, one player in for next cycle, then no players in for three cycles. One player out for each of the five cycles.

For trials 2, 3, and 4 follow the same inquiry process as you did in the first trial. Ask your participants to predict what will happen to the number of players in the stock. Conduct the trial, and then discuss the relation of predicted and actual results.

Here is a summary of the results you will experience over the four trials:

Trial 1: If the inflow is greater than the outflow, the contents in the stock will increase. In this trial we start with 6 in the tub and add 2 each cycle; outflow is 0. The stock content increase goes: 6, 8, 10, 12, 14, 16.

Trial 2: Even if the inflow is made smaller, the contents can continue to get bigger if the inflow is greater than the outflow. Start with 6. For two cycles, the inflow is 2; for the last three cycles, it is

1. For all cycles, outflow is 0. The stock content increase goes: 6, 8, 10, 11, 12, 13.

Trial 3: If the inflow is the same as the outflow, the amount in the stock will remain the same. Start with 6. Inflow is 2 and outflow is 2 for all five cycles. The stock content goes: 6, 6, 6, 6, 6, 6.

Trial 4: Even if the inflow goes to zero, the contents can remain large for a long time. Start with 6. Inflow is 2, 1, 0, 0, 0. Outflow is 1 for all five cycles. The stock content goes: 6, 7, 7, 6, 5, 4.

Explain: *"You have just experienced a stock/flow structure common to a wide variety of systems. An amount of something that increases or decreases over time—trees, fish, people, goods, clutter—is a stock. A stock is not changed directly. Its size is controlled only by altering flows in and out of the stock. This activity is very simple, but it illustrates an extremely important fact about climate change dynamics."*

Note: The amount of GHG lost from the atmosphere each year depends partially on how much is in it. In this exercise this feedback is excluded to simplify the mechanics of the game.

Debrief

Ask, *"In what ways is the atmosphere like a bathtub?*

"If you are thinking that the atmosphere accumulates carbon dioxide and other greenhouse gases the way a bathtub accumulates water, you are right.

"Most climatologists agree that humans are putting greenhouse gases into the atmosphere at almost twice the rate that natural processes remove them. The flow into the tub is twice the rate of flow out.

"When you are already emitting GHGs at twice the acceptable rate, slowing the rate of growth will still lead to increasing the overall levels, not to reductions.

"Even if policies to mitigate climate change caused greenhouse gas emissions to fall, atmospheric greenhouse gas concentrations would continue to rise until emissions have fallen to the removal rate."

For your debriefing, you might want to use some current actual figures to indicate current levels of CO_2 in the atmosphere and its annual rise. Show current proposals to cut emissions. Ask how these two figures relate. Most government and intergovernmental proposals to cut emissions are not aggressive enough to stop the growth in GHG concentrations.

5.

Biodiversity Game

You can't change only one thing.

Quotes

Scientists themselves readily admit that they do not fully understand the consequences of our many-faceted assault upon the interwoven fabric of atmosphere, water, land, and life in all its biological diversity.

—Charles, Prince of Wales

The International Union for Conservation of Nature Red List of Threatened Species shows that 20,000 out of the 64,000 assessed species are threatened with extinction.

—International Union for Conservation of Nature, 2012 Red List

What breaks in a moment may take years to mend.

—Swedish proverb

Climate Link

Though the science of biodiversity is still quite primitive, we know that species are disappearing on this planet at an accelerating rate. And there are several reasons to suppose that the rising pace of climate change will increase the speed of the extinctions.

Many mechanisms pose challenges to biodiversity. Animals are adapted to specific temperature ranges and precipitation regimes. As these change, the suitability of habitats shifts. Often it will not be possible for a species to migrate to new regions that are still hospitable quickly enough to avoid extinction. One extreme case is when species respond to rising temperatures by moving gradually up a mountain to higher elevations. When they get to the top of the mountain, there is no adjacent place remaining to go in response to further temperature rises, and they die out. Shifting cultivation zones will induce migration of human populations into regions that are now sufficiently under-populated to support wild species. The wild species will be crowded out. Pests and predators will move into regions that are now occupied by species that do not have appropriate defenses.

Each species depends on others for food, pollination, shelter, protection from predators, and for other needs. When one species disappears, it will certainly affect others in ways we cannot yet precisely predict.

About This Game

This exercise is designed simply to reinforce the idea that species do not exist in isolation. When one species disappears, it will certainly cause the elimination of others that are interrelated with it.[5]

This exercise is based on a master triangle that is divided into nine smaller, equally sized triangles.

It offers an analogy for biodiversity, though the interrelation among adjacent triangles is obviously completely different from the interactions among interdependent species. The process of removing small triangles from a larger cluster provides no detailed information about possible future changes in the number of species. The exercise merely offers a shared experience of one important point and can facilitate useful discussion about the nonlinearity of species extinction.

To Run This Exercise

Number of People

This is a mass game that can be conducted with any number of participants, from one to thousands. If you have a small group and sufficient time, you could give each participant a piece of paper containing a picture of the master triangle. Then ask them to use their own pens or pencils to determine what happens to the number of triangles when different lines are eliminated. With a larger group and less time, you can show them the questions and the answers on slides, pausing long enough to let each of your participants develop a preliminary answer to each of your questions.

Time

The time required depends on the mode you use for running the game. It will range from a few minutes up to twenty minutes.

Space

This exercise is suitable for a large seated audience. It is only necessary that every participant can see and hear you.

Equipment

Paper and pencil for each participant.

Setup

Draw, print, or otherwise recreate the triangle illustrations shown in this exercise. It is simplest to project a slide of each figure on a screen.

Instructions and Script

Conduct the exercise without first explaining its purpose. It is important to get everyone to participate in the process and to give each of them time enough to form an answer to each of the questions you pose.

Step 1: Begin by putting up the master triangle on a flip chart or projection screen.

"Please look at this figure. How many different triangles are formed when nine small triangles are connected as shown here?" Emphasize that all members of the audience should form their own answers and give them a minute to calculate.

"That form contains thirteen different triangles." Show them the answer. *"There are nine small triangles, three larger ones, and the whole."*

Step 2: Ask: *"How many triangles will be left if you take away one of the small ones?"* Give them two minutes to decide on their answer.

"In fact it is impossible to remove only one triangle. If you remove one of the smallest triangles from the thirteen, because their borders go with them, three or five other ones also disappear, depending on which one you remove. You will be left with a figure that contains only seven or nine triangles."

Debrief

Because they are connected to each other, you cannot reduce the number of triangles by only one. Taking away one triangle also eliminates three or five others. Species are more connected to each other than these triangles. Whenever we lose one species, we will inevitably lose others.

Here are a few questions you could ask during the debriefing:

- *"What are the mechanisms through which we could expect that climate change will cause species extinction?"*

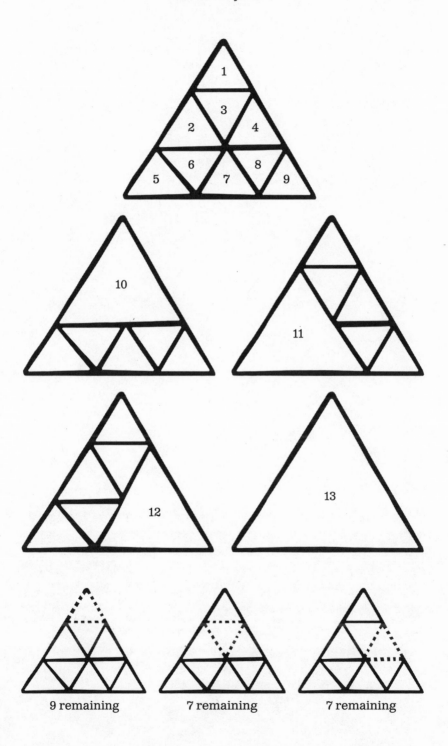

9 remaining   7 remaining   7 remaining

- *"What are some species that are threatened by these mechanisms?"*
- *"In what ways could the elimination of one species threaten other species?"*
- *"What can we do now to preserve species during a period of rapid climate change?"*

6.

Circles in the Air

Our perspective affects the actions we take in complex systems.

Quotes

There now needs to be a change of perspective to take this [climate change] as an opportunity (rather than a burden) . . . and that is the key to green growth.

—Thomas Heller, expert in international law

Some countries of the world are like people fighting on a large boat. In the middle of their battle over how to catch the biggest fish, they look up and realize that the fine boat is sinking, and everyone is going down. Their next fight will be for basic survival, and they will need to rely on one another, floating far from shore in the angry sea.

—Julie A. Barnes, founder, The Amazing Seed Foundation

Climate Link

When it comes to any particularly complex problem, we all have a propensity to look at our part of the system or, worse, to consider ourselves outside of the system and to place the blame for the problem on someone else or some other group. This is particularly true for the issue of climate change.

Circles in the Air is a terrific, experiential means to explore different perspectives related to climate change. For example, scientists say that it is urgent that we reduce carbon emissions, whereas the general public may think (erroneously) that no action is necessary because climate change is easily reversible. These are wildly different perspectives. An advocacy group may have a strong opinion about who or what needs to change to address global warming—whether government, business, North or South, one country or another—that is likely to be very different from the opinions expressed by a global corporation.

This exercise illustrates how our perspectives affect the actions we take within and about complex systems. It subtly focuses a group's attention (in a fun and nonthreatening way) on thinking about its own thinking.

About This Game

This exercise works on many levels. It exposes the natural tendency to see one's self, group, organization, or even country as outside the system, and to perceive causes of problems as "out there." As people go through the exercise, they quickly discover how their perspective can vary based on the view they have of the same system. They also discover that by changing their vantage point, either mentally or physically, they can potentially discover new insights and new leverage points.

When used to explore the theme of climate change, this exercise can help individuals and groups:

- Develop greater awareness of "the enemy is out there" syndrome;
- Identify the different viewpoints related to climate change and explore the possibility that our viewpoint depends upon where we sit; and
- Set the context for discussing the concept of underlying structure.

To Run This Exercise

Number of People

This is a mass game. It will accommodate any number of people. Participants can either be sitting down or standing up.

Time

Two to ten minutes (depending on the length of the debrief).

Space

Just enough room to be able to point a thumb in the air.

Equipment

None.

Setup

None.

Instructions and Script

Step 1: Ask your participants to hold one of their thumbs up in the air above their heads and to keep their thumbs always pointing up.

Step 2: Have participants use their thumbs to draw a circle in the air in a clockwise direction, always keeping their thumbs pointed upward.

(You can demonstrate this.) Say, *"Your goal is to move your hand clockwise. Always keep your thumb pointing up. Do not stop the rotation once it starts and do not change the direction."* Tell them to continue drawing the circle and looking up at the tops of their thumbs.

Step 3: *"Now, continuing to draw the circle clockwise, slowly bring your hand down a few inches at a time until it is in front of your face. Continue to circle your thumb and slowly bring it down to the level of your waist until you are looking down on top of it. Continue to trace out the circle while looking down on it."*

Step 4: Ask the group, *"What direction is your thumb moving?"* It will be a counterclockwise direction at this point.

Note: You will find that some people lose the integrity of the circle as they bring their thumbs down, swishing their hands back and forth in a straight line. If you notice this, suggest starting over and encourage them to practice drawing a round circle on the ceiling before moving their thumbs down. You may also notice that some people change the direction of the circle as they lower their arms. Simply bring this to their attention and demonstrate, once again, how to continue to draw the circle in the same direction as they lower their forearms.

Debrief

To begin, simply ask: *"What happened?"* The initial responses tend to range from the insightful ("My perspective changed.") to the humorous ("You tricked me!").

After people have had a chance to try it again, most of them will see that what changed as they brought their thumbs down was not the direction their thumbs were moving but their perspective or vantage point. The debrief can go in any number of directions from here.

Acknowledge the looks of astonishment and surprise. Then weave that reaction into a conversation about the potential of changing perspectives to achieve a greater understanding of complex systems, in particular climate change.

As part of your debrief, here are several questions you may ask:

- *"What was your initial reaction?"*
- *"Do you remember the language you used to describe what happened?"*
- *"Did your immediate reactions provide any insight into your own process of forming assumptions or into your own reaction when you are confronted with a situation that is confounding or puzzling?"*
- *"What does this have to do with climate change?"*
- *"How is it that we may all be looking at the climate change system from one perspective? Can you find ways to look at it from multiple perspectives?"*

You can use this exercise to bring a group's attention to the various perspectives they hold.

Holding your thumb up straight in the air, say: *"The clockwise direction represents countries with heavy emissions and their perspective on climate change."*

Then lowering your thumb: *"The counterclockwise direction represents countries contributing to climate change debate with adaptation messages."*

or:

Holding your thumb up straight in the air, say: *"Scientists have concluded we must reduce carbon emissions immediately!"*

Then lowering your thumb: *"From this perspective, that of the general public, it looks as if no action is necessary. Technology will save the day and reverse the impacts of climate change."*

The structure of the climate change system is the same. What is different is the level of perspective at which we are looking at the systems.

You may also use this quote from system dynamicist and author Donella Meadows to spark a conversation about level of perspective as it relates to climate change:

"How is it that one way of seeing the world becomes so widely shared that institutions, technologies, production systems, buildings, cities become shaped around that way of seeing? How do systems create cultures? How do cultures create systems?"[6]

7.

Frames

**To obtain consensus be clear about
the mental framework you are using.**

Quotes

When you're walking along a tricky, curving, unknown, surprising, obstacle-strewn path, you'd be a fool to keep your head down and look just at the next step in front of you. You'd be equally a fool just to peer far ahead and never notice what's immediately under your feet. You need to be watching both the short and the long term—the whole system.

—Donella Meadows, environmental leader

Sometimes I'll peek out from underneath the focusing cloth and just look around the edges of the frame that I'm not seeing, see if there's something that should be adjusted in terms of changing the camera position.

—John Sexton, president, New York University

There is more light than can be seen through the window.

—Russian proverb

Climate Link

Great folk wisdom is captured in a body of folktales about an itinerant Sufi preacher, Nasreddin. In one story Nasreddin frantically searches for something under the light of a lamppost in the dusty street outside his home. A kind neighbor comes by and asks, "Mulla, what have you lost?" Nasreddin replies, "I have lost my keys." The neighbor, being the good person he is, gets down on his hands and knees and begins to search with Nasreddin through the dust. After a long time, the neighbor says to Nasreddin, "Mulla, are you certain you lost your keys here in the street?" "Oh, no!" says Nasreddin, "I lost them in the house." "If you lost them in the house," says the neighbor, "then why are we looking for them under this lamppost?" "The light is better here," Nasreddin replies.

Like Nasreddin, we often search where the light is better. When we have problems, we look for their roots in the data that are most easily accessible to us. In the language of systems thinking, we say that people look for proximate causes—that is, causes that are close to the problematic symptoms in both time and space. Yet the causes of difficult behaviors typically lie far away from the physical location where the actual behavior is observed and far back in the past.

Climate change provides ample opportunities to observe both of these facts. We know that the greenhouse gas emissions on one continent can have dramatic effects on the climate patterns in others—which is why the world is concerned about large-scale projects to build coal-fired power stations in places like China, India, Germany, and the United States. We also know there are delays in these complex systems, and it is hard to convince people to act now on something that will result in changes many decades in the future.

The link from CO_2 emissions to climate change is long and full of delays. Even if all CO_2 emissions from human sources were eliminated tomorrow, the average global temperature would continue to increase for decades or even centuries, from all the greenhouse gases that society has already emitted. Therefore actions being promoted and decisions that need to be made are oriented to longer time frames than many people are accustomed to considering or even willing to consider.

The search for some basis to choose among alternative approaches or policies tends to concentrate where the light is better. Preference is given to policies that provide benefits, here, and soon. But the important consequences of actions that affect climate will typically occur far away from the location of the action and years—or decades—into the future. Debates about the appropriate policy to adopt in relation to CO_2 emissions or the exploitation of oil reserves or the protection of endangered species all illustrate this tendency.

About This Game

To understand the climate predicament and find more effective solutions, it is often essential for us to reframe our dilemmas and redefine their boundaries consciously and deliberately. The primary challenge to this is that we are often unaware of how we are framing a problem. If we cannot see the current frame in which we have cast an issue, we find it difficult to change that frame, even if the old perspective is keeping us from understanding and addressing our troubles. When we are under pressure, we tend to focus even more on what we're seeing through our frame and pay even less attention to the frame itself.

This exercise shows how to *have* our frames as opposed to being *had by them*. It helps illustrate the impact of choosing different frames or points of view when defining a problem and seeking solutions. It can encourage participants to try on different perspectives when diagnosing problematic system behavior or when designing ideal systems.

It reveals why two different, intelligent people of goodwill can have opposing views about climate change—one thinking it is a serious problem and the other believing it does not even exist.

This exercise can also help people develop greater openness to experimentation with different time horizons when defining a problem or solution. It can help support more robust objectivity about paradigms and increase the willingness to realize that one's view of reality may not match others' view of it and may not be the most useful way to view it.

Most societies have several common frames through which people view reality. These frames include religious beliefs, economic theory, natural science, or political ideology. These frames are paradigms—filters that direct one's attention to specific forms of data, predispose one to specific theories of causality, and focus one on established kinds of problems and policies. All paradigms have three key aspects:

1. An implicit time horizon; that is, the length of time over which we consider information about the issue;
2. A geographical boundary that defines where we look for costs and benefits of alternative policy options;
3. Causal links that are presumed to be important. For instance, many economists disregard feedback from the environment in advocating their favorite policies. Many environmentalists, for their part, disregard the impact of the price system when arguing their own viewpoints.

Efforts to solve problems typically start without any effort to determine which frame might be most useful. A research experiment conducted some years ago by NASA offers an apt example. In 1978 NASA launched the satellite *Nimbus 7* into the stratosphere to gather long-term data on significant atmospheric changes high above the earth. However, the people who designed the experiment were working under an unexamined paradigm. They assumed that they would not have to measure ozone concentrations, because they believed that such concentrations did not change. Consequently, they programmed the computers on board the satellite to ignore information about ozone levels. Therefore, although the satellite did sense changes in ozone levels, the data were not transmitted back to NASA. If the experiment designers had operated from a different paradigm, we all would have learned much earlier about the grave damage chlorinated hydrocarbon chemicals were causing to the earth's ozone layer.[7]

Frames become especially important during times of major change in the world around us. If we are not in the habit of changing our frames, we may inadvertently maintain an old one long after it is no longer relevant. Professor of public administration and policy George Richardson points out two kinds of frames or boundaries that are particularly interesting to explore within the context of climate change learning: geographical (or spatial) frames and temporal (or time horizon) frames.

1. **Geographic, or spatial, frame.** This boundary defines the physical area over which we think people, organizations, and natural systems will be affected by the actions we take. If we adopt a narrow geographic frame, we will pay less attention to consequences that occur "over there" than we do to events that occur in our own backyard. Some nations oppose climate change initiatives because they believe that climate change will be mainly beneficial within their own boundaries. Those countries are operating from a geographic frame that ignores damaging effects of their actions outside their own borders.

2. **Temporal, or time-horizon, frame.** This boundary defines the interval of time over which we care about the costs, benefits, or results of the actions we are considering—for example, one hour, one week, one year, ten years, 100 years. Almost everyone gives less attention to costs and benefits that will occur in the far future than they do to those that are immediate. Economists have even coined the term "discount rate" to designate the degree to which we devalue, or disregard, future costs in favor of current benefits. If you have a high discount rate, you ignore information about consequences that will manifest themselves more than a few years into the future.

This temporal frame is particularly evident and damaging among elected politicians with short terms of office, but everyone suffers from it. If people felt the effects of their greenhouse gas emissions immediately and personally, they might choose to undertake fewer emission-producing activities. But in this case the consequences lie decades in the future, so most people ignore them and opt for the immediate pleasure of their current lifestyles. All the important activities that contribute to climate change give pleasure, prestige, or profit to at least some people and organizations in the short term. So long as they have a short-term time frame, they will oppose effective efforts to reduce greenhouse gas emissions.

The time-horizon boundary also has a moral dimension, per Richardson. For example, if you are thinking about energy in a one-year time frame, you may focus on price and supply. But if you take a 200-year time frame, you cannot afford to ignore issues of climate change and unequal quality of life among generations.

To Run This Exercise

Number of People

This is a mass game. It can be used with any number of participants.

Time

Most people use this exercise as a simple illustration within a longer and more substantive discussion of conceptual frames or boundaries. In that case you can present it in five minutes. If you wish to use this exercise as the basis for a more extensive discussion, allow fifteen to thirty minutes.

Space

Participants simply sit in place. The exercise does require everyone to look at you from a distance of at least 6 feet (2 m) away. The exercise is described as if you are standing in front of the audience, though you could adapt it for use in a circle.

Equipment

None.

Setup

None.

Instructions and Script

Ask each participant to use a thumb and forefinger touching at the tips to create a ring to look through. This will be their "frame."

Spatial Framing

Step 1: Divide the room into two halves down the middle. Hold your arms and hands outstretched from your sides. Ask all participants to hold their frames—the ring each has formed with a finger and thumb—out at arm's length directly in front of them toward you and to look through the ring at one of your hands. Ask those to your right to focus on your right hand; ask those on your left to focus on your left hand.

At this point you are holding both arms outstretched from your sides. Your right hand is thumb up. Your left hand is thumb down. Ask all participants to look through the frames they have created and focus on the indicated hand.

Step 2: Give the following instructions, pausing for ten to twenty seconds after each so participants have time to ponder their response. *"Everyone who thinks my thumb is pointing up, raise your hand."* (Pause.) *"Everyone who thinks my thumb is pointing down, raise your hand."* (Pause.) Typically, those on one side of the room will disagree with those on the other.

Step 3: Now ask the participants to bring their frames as close as they can to their eyes while keeping the same hand, either your right or your left, centered in their frames. Again instruct them to raise their hands if they think your thumb is pointing up, pause, then ask those who think your thumb is pointing down to raise their hands and pause again. This time you usually will get everyone in the room to raise their hands twice. When the frame is close to their eyes, they have a much wider field of view. They see much more of reality; they normally can see both of your thumbs. Often people do not disagree because they have a different reality; they disagree because their frames cause them to look at different parts of reality.

Temporal Framing

Step 1: Remind participants that the room is divided into two halves—left and right. Once again ask participants to look through

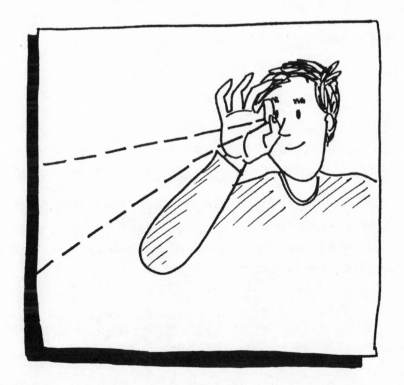

their personal finger frames, this time holding the frames as close to their eyes as they can and all looking at your right hand. All should start looking when you say "Go." After five seconds those in the left half of the room will quit looking through their frames; those on the right half will continue to look through their frames at your right hand for ten seconds more—a total of fifteen seconds.

Step 2: Hold out your right hand with only your thumb extended up and all your fingers bunched into a fist.

Ask everyone to begin their observations. Start your timing. After five seconds announce, *"Those in the left half of the room may stop observing now."*

After five seconds more (after ten seconds have elapsed altogether) extend all five fingers of your right hand.

Five seconds later, after a total of fifteen seconds, tell those in the right half of the room, *"You may stop observing now."*

Debrief

Here are some suggested debrief questions:

- *"Who thought the number of extended fingers on my hand changed while they were watching? Raise your hand?"*
- *"Some thought my hand changed; some did not. What is the truth? How can reasonable people disagree on such a simple question?"*
- *"What is the relation of this exercise to observations about climate change?"*
- *"What period of time is implicit in the data used to think about climate change?"*
- *"How is it possible to know if the view of the system is long enough to detect important changes?"*
- *"How could the length of time implicit in people's discussion about climate change be increased?"*

8.

Group Juggle

Adding one more apparently minor problem
can sometimes collapse the whole system.

Quotes

A lack of appreciation for what exponential increase really means
leads society to be disastrously sluggish in acting on critical issues.

—Thomas Lovejoy, conservation biologist

Global warming, with all of its attendant changes, is the first clear mega-symptom revealing that we are now rapidly approaching many tipping points where overshoot and collapse are occurring.

—Dave Steffenson, acting director, Wisconsin Interfaith Climate and Energy Campaign

The last drop makes the cup run over.

—English proverb

Climate Link

People have become accustomed to the idea that problems grow slowly in number and that it will usually be possible to deal with each as it arises. This reveals a deep misunderstanding about the role of exponential growth and the capacity of systems to cope. The error persists even among highly educated adults. Why? Most adults today were not explicitly taught skills related to seeing systems of multiple causes, effects, and unintended impacts. These are the skills needed to navigate issues of global impact such as climate change, interdependent financial markets, and biodiversity loss.

In the early decades of the industrial era, the amounts of GHG being added to the environment by human activities were low. So even though they were increasing exponentially, they remained small. This generated new problems only very gradually, and humanity could cope. But we are entering an era where that steady exponential growth is being applied to larger and larger emission streams. The problems are mounting. And they have the capacity to overwhelm society's adaptive powers. Rather than a slow deterioration, we may see sudden collapse.

About This Game

Understanding the systemic nature of challenges such as climate change can be gained through direct participation and by reflection on personal experience. Group Juggle is a valuable tool for this purpose. It is also fun. With balls or other objects that can be tossed, the game propels people up and out of their chairs and gets their blood circulating. This exercise can generate some real laughs, and it almost never fails to reveal profound new insights.

Specifically, you can use this exercise to:

- Illustrate the way a simple set of rules can produce complex behavior;
- Give participants the experience of being part of a system in which the pattern of behavior rapidly shifts from coping to collapse;
- Break down the formal, social barriers that exist in a workshop when its members first assemble;
- Develop awareness that different groups will behave in similar ways when they are immersed in systems that employ the same rules.

To Run This Exercise

Number of People

This is a participation game. Each session involves a team of fifteen to twenty people. If you have more than twenty players, split the group into smaller teams. Lead each team through the exercise in sequence.

Time

Fifteen to sixty minutes, depending on the number of lessons you want to convey, the level of participants' sophistication, and the length of the debrief.

Space

Enough open space to let team members stand 3 to 5 feet (1 to 1.5 m) apart in a circle. Objects will be tossed into the air, so the space needs a ceiling height of not less than 8 feet (2.4 m).

Equipment

A flip chart or white board;

One ball or other tossable object per participant (such as a tennis ball or softball);

A box, shopping bag, wastebasket, or other container to hold the balls.

Setup

Put a chair within your reaching distance. Put the balls into the container, then put the container on the chair so it is easy to reach the balls without bending down.

Things to Consider

If a participant is unable to catch or toss a ball or other soft object, you can ask that person to play the role of a process observer who provides commentary and feedback to the group during the exercise debrief.

Instructions and Script

Step 1: Arrange your team or teams. If your group consists of more than twenty people, divide them into smaller teams so no team has more than twenty members. Designate one team to play first (the beginning team) and the other team(s) to serve as observers. Stand with the beginning team in a circle. Ensure that members of the observing team(s) stand far enough away from the circle that they will not impede people who will leave the circle to retrieve dropped balls. However, observers should stand close enough to watch the exercise unfold.

Step 2: Establish the throwing order. All group members start by holding their hands out in front of them at waist level, elbows bent. You, the facilitator, throw the ball to someone. After that person catches the ball, he or she looks for receivers whose hands are still raised at waist level, throws it to one of those people, then lowers his or her own hands. The person who just caught the ball throws it to someone else and then lowers his or her hands, too. Encourage people not to throw the ball to someone right next to them but to someone across the circle who still has his or her hands raised. Continue until the ball has been thrown once to everyone and everyone's hands are lowered.

The goal during this part of the exercise is accuracy, not speed. Underhand throws are easier to catch. If anyone drops the ball, ask him or her to retrieve it and resume the sequence of throwing.

Ask each team member to remember who caught the ball when he or she threw it. Each person will always have one designated catcher during the game. During this initial throwing process, ensure that no one gets the ball more than once.

The last person in the initial sequence, when everyone else has dropped their hands, should throw to the person who received the ball from you.

When that person has received the ball for the second time (the first time being from you), stop this process of establishing the throwing order. In the actual exercise the person who initially caught the ball from you will be the designated catcher of the last person in the team to get the ball. Thus, once you throw a ball into the circle, it should continue to circulate around the group indefinitely unless it is dropped.

Get the ball that was used for this trial run and put it in the container.

Step 3: Explain the rules of the game: *"Your team's goal is to keep as many balls as possible in the air at the same time. Do this by continually catching balls from your designated thrower and then throwing them to your designated catcher."*

And explain what you will do: *"I can throw to anyone I see who is not currently holding a ball. We will start the game slowly. But as I see you successfully keeping more and more balls in the air, I will throw more and more balls in."* Make certain that everyone has heard this statement; it forms the basis for the overshoot behavior in the early phase of the game.

If there are observers, ask them to gather data about the number of balls in the air over the course of the exercise.

Step 4: Who is each player's designated catcher? Test to be sure everyone remembers. Ask team members to simulate throwing the ball by pointing in sequence to their designated catchers. You start the process by pointing to the person who received the ball from you. That person in turn points to his or her designated catcher. Continue in this way around the circle. If anyone has forgotten the identity of his or her catcher, have the group figure it out. In rare cases you may need to throw the ball around again to clarify the sequence or establish a new one.

Step 5: Carry out the game. Throw a ball to the first person you threw to during the sequence-determination round. As team members start passing the ball around according to the established sequence, wait five seconds. Then throw in another ball. Wait three seconds. Then throw in more and more balls. Soon you will overwhelm the capacity of the group to maintain the balls in the air. When people start dropping balls, usually after there are ten to fifteen balls in play, loudly urge the players to retrieve them. To provide even more distraction and give the group a few laughs, you can throw in a rubber chicken or some other bizarre but harmless object. As the chaos grows, start throwing balls in rapidly to many different people, even if they are clearly not ready to catch them. Then call out: *"Okay, stop!"*

Step 6: Switch teams. If you divided the group into different teams, go through these steps again with the second team, asking the other teams to observe the number of balls in the air. After all the teams have participated in the game, move to the flip chart for the debrief.

Debrief

This exercise is rich in content. The recommended debriefing process is sketched below. Guide your group through this sequence in a leisurely fashion, show them the relevant illustrations, and give players plenty of opportunities for questions and discussion.

- First, ask questions that give participants the opportunity to share their general impressions of the activity. Allow them to share their observations about the events of the game: *"What happened? How did the behavior of the group change over the course of the exercise?"*

Now move into a discussion about behavior over time, a key systems concept that the game illustrates. Explain: *"To take a deeper look at what happened, let's look at the patterns we observed. At first, I threw in just a few balls, and you did a perfect job of juggling them. Then I threw in a few more, and still you did a great job. That initial behavior, competence, is shown as mode 1 in the illustration."* Refer to the illustration below, drawn on your flip chart or white board.

Explain: *"But you did not continue coping. Consider the pattern of balls in the air over the remainder of the exercise."*

Explain: *"Once you began to drop the balls, I quit throwing any more into the circle. Notice that the pattern of behavior over time went through three modes:*

- *First was competence—coping. As I threw in more balls, you caught more.*
- *The second mode was limits to growth. I threw even more, and initially you caught even more. But then your group reached its capacity to process the balls. As I threw in more, you no longer caught more. What happened to the extra balls? They were dropping to the floor and being retrieved.*

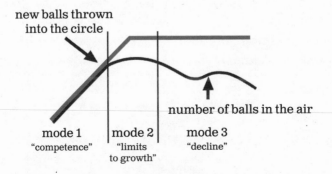

- *The third mode—decline—came soon after that. Though I was not throwing in more balls, your capacity to keep them in the air went down. Because participants were diverted to retrieve the fallen balls, your group's juggling capacity declined."*

Ask: *"Why did this happen, and what could you have done to perform better?"* Give players time to discuss and reply.
Typical responses:

- "At first it was easy, since there was lots of time to get ready for the next ball. Then they started coming faster, and I started to drop them."
- "I think if we practiced catching the ball, we could do better."
- "I think if all the balls had been the same, we would not have dropped so many."
- "Maybe we could have beat a drum or created a rhythm in some other way. Then we could all have thrown our balls at the same time."

Point out, *"This game is a useful metaphor for society's capacity for dealing with the problems from climate change."*

Ask your participants to imagine how the Group Juggle illustrates behavior related to climate change dynamics. You may ask, for example: *"How might the three behavior patterns we experienced in Group Juggle play out when we think of the consequences of climate change?"*

Pause for discussion, then explain: *"Here is one example: if you are confronted with one problem, for example, flooding, then you deal with that problem. However, if the number of problems gets bigger, floods and storms and crop loss and pests, then the system may become overwhelmed and collapse."*

9.

Hands Down

When trying to understand a complex situation,
don't limit your focus to where the action is.

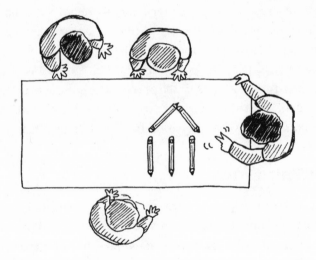

Quotes

Out of sight, out of mind.

—English proverb

We have a system of national accounting that bears no resemblance
to the national economy whatsoever, for it is not the record of our life
at home but the fever chart of our consumption.

—Wendell Berry, novelist and environmental activist

Climb mountains to see the lowland.

—Chinese proverb

Climate Link

The data available to policy makers inform our economic system and our technology. There is an automatic tendency to look within the spheres of economic action and technological change for the solutions to problems. The debate over actions to counter climate change focuses on economic initiatives (such as carbon taxes, electricity pricing, imposition of levies that reflect external costs) or on technical change (improved energy efficiency, better ways to remove carbon dioxide from emissions, or enhanced solar power devices, for instance). There are fewer sources of data about population and lifestyles, so these spheres remain outside the policy debate. This exercise is designed to show participants how one's frame of reference influences the solutions one considers and to encourage participants in the climate debate to be more creative in where they look for data to guide their actions.

About This Game

This exercise can help raise awareness of our unexamined assumptions. It can encourage participants to slow down and look at every assumption, especially the unconscious ones, in order to explore the double-edged nature of mental models that enable us to function in the world but also often act as blinders. The most-needed and most often underdeveloped skill is using peripheral vision, going wide. The wider our perspective, the more data we can take in and the more possibilities exist for effective action.

To Run This Exercise

Number of People

This is a mass exercise. Any number may play as long as all participants can see what you are doing. If you have a small group—eight or fewer—you can simply use your hands and five or six pens. For larger groups, use a flip chart.

Time

Five to fifteen minutes, depending on the size of the group and the extent of the debrief.

Space

Enough room so the group can assemble around you or observe a flip chart. A seated audience can easily play this game.

Equipment

For small groups of eight or fewer:
Six or seven pens
For larger groups:
Flip chart;
Thick marker that is easy to see from a distance.

Setup

A minimum amount of setup is best. The more spontaneously you can pick up your pens or turn over the flip chart sheet and draw on the frame, the better.

Instructions and Script for Small Groups

Step 1: Explain, *"I am going to use a secret code to show you numbers between one and five, only integers. Your goal is to decipher my code*

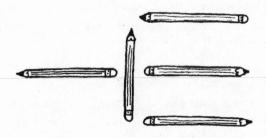

and understand each number I show you. Of course, you will make mistakes the first few times until you understand the code."

Arrange the pens on the floor or on a table, in full view of the group. Be a little fussy as you put down the pens, as if you are arranging them in some particular pattern.

Step 2: After you've arranged the pens in front of you, place your idle hand flat against the surface with all five fingers outstretched. Ask the group: *"What number am I showing?"* Pause.

Step 3: Individuals in the group will share numbers. Many will simply say the number of pens showing. Others will try to figure out the "code," interpreting the arrangement of pens to mean some particular number. Others may guess the right answer, in this case five. The number of fingers you show on your idle hand is the answer to "What number am I showing?" Try this several times, varying the number of fingers you show on your idle hand. Complete three or four rounds; then proceed to the debrief.

Instructions and Script for Larger Groups

Step 1: Draw a large rectangle in the upper right-hand corner of a flip chart. Draw a first set of three or four simple marks within the rectangle. As you do this, act as if you are arranging them in some particular pattern.

Step 2: As you are drawing the symbols inside the rectangle *casually* place your idle hand flat against the drawing surface with all five fingers outstretched. Ask the group: *"What number am I showing?"* Pause. *"The number is five. Who guessed five?"* Normally about one-fifth of the people will raise their hands.

"Let us try again." Cross out or erase the old pattern, then draw a new pattern using four or five marks.

Casually remove your idle hand and then replace it on the surface, outside the rectangle. But this time only three fingers are outstretched.

Ask, *"What number am I showing?"* Pause. *"The number is three. Who guessed?"* Go through this process one more time with two fingers. When you announce that the answer is two, there will start to be some frustration in the audience.

Step 3: Stop drawing the symbols inside the rectangle and proceed quickly. Put your idle hand on the surface with only your thumb out-stretched: *"Now I am showing one."* Change your hand to show four fingers: *"Now I am showing four."* Do this several more times, each time showing a different number of fingers, until people understand your code. Finish up with: *"Were you looking for information inside the rectangle? That has nothing to do with the true number. The code is very simple. But as long as you focus inside the rectangle, you will never, ever understand it."*

Debrief

Without a thoughtful debrief, this exercise could easily fall into the gotcha category and leave participants feeling frustrated and even manipulated. Be sure that they understand that even intelligent, careful people will be fooled if their attention can be focused on the wrong information. If you've chosen to draw a rectangle, you may note that in this activity it is particularly useful to use the term "think outside the box."

Here are some suggested debrief questions:

- *"What frame do we use to define the relevant data for discussions about climate change?"*
- *"Who determines that we should use that frame?"*
- *"What would be a more useful boundary for us to use in considering information about climate?"*
- *"How could we change the boundaries implicit in people's discussion about climate change?"*
- *"Governments delight in using gross national product (GNP) data as a frame for reporting on the results of their decisions.*

However, GNP calculations lump together activities that are beneficial with those that are harmful. And GNP calculations ignore the future costs that will result from current actions. How could the GNP frame be modified to help avoid activities that will damage the climate?"

10.

Harvest

Over the long term, individuals often get more
from cooperation than from competition.

Quotes

If you cannot catch a fish, do not blame the sea.

—Greek proverb

One reason it is so hard to slash carbon emissions is that climate change occurs globally. The countries that produce the most greenhouse gas all need to take action to fix the problem. That raises a classic economic dilemma called the tragedy of the commons.

—David Kestenbaum, science correspondent
for National Public Radio

A courtyard common to all will be swept by none.

—Chinese proverb

Climate Link

Our "commons" are those resources such as air, water, land, highways, fisheries, energy, and minerals upon which we all depend and for which we are all responsible. When it comes to our commons, we may find ourselves acting in surprisingly selfish ways. In 1968 the ecologist Garrett Hardin popularized a phrase for this strange behavior. He called it "the tragedy of the commons."

We need common resources—air, water, fish—to promote our own well-being. Without some collective agreement about how that common resource will be managed, however, these commons typically become overused, resulting in a collapse of the very resources on which our well-being depends. The global atmosphere is a commons. When countries act in their own interests and emit high levels of CO_2 into the air, they experience the immediate gains; greater economic growth, for instance. But they do not experience the future and distant losses related to global warming. If just one country emitted high levels of CO_2, there might be little to no negative consequences. But when many countries emit high levels of CO_2, everyone suffers.

The Harvest game offers participants a chance to manage a commons sustainably. Using an imaginary fishery, participants can experiment with different modes of cooperation and partnership to avoid a tragedy of the commons.

About This Game

Harvest provides an experiential means for participants to see the long-term consequences of using a limited renewable resource to maximize their own short-term benefits. This game also reveals what can happen when a few players (whether individuals, organizations, or countries) dominate a system to the detriment of the collective good.[8] This exercise can be used to explore the archetype of the tragedy of the commons and the related phenomenon of worse before better.[9] In worse-before-better scenarios, the actions required to produce fundamental, long-term solutions often make the situation seem worse in the short run. The long-term results can prove tragic when politicians or economists persist in looking only at the short-term indicators of success as they select policies. This fact has been graphically illustrated in many sectors of society, but especially in the flagrant overuse of natural resources such as fishing grounds. In many areas overfishing has enormously reduced the fish populations' ability to regenerate.

To keep using a renewable resource in the long term, it is often necessary to accept a short-term reduction in its harvest. Implementing sustainable-use policies requires understanding the system's long-term dynamics, valuing long-term welfare above short-term gain, and trusting others to observe short-term constraints. Harvest gives groups the opportunity to practice all of these principles.

To Run This Exercise

Number of People

The exercise requires a minimum of four participants. It works well with groups up to about forty.

Time

Thirty to fifty minutes.

Space

You will need to select a space that accommodates three kinds of activity. First, you will introduce the game to the whole group of participants. Second, they will play the game. Third, you will lead participants through a debriefing conversation. It is most convenient to conduct the first and third of these activities using a flip chart placed in front of enough chairs to seat your entire audience. The second needs a room that allows people to break into teams of two to six people. These small teams should be able to sit or stand far enough apart that they do not overhear each other's conversations.

Equipment

One large container per game to represent the ocean. A coffee can works well.

Three hundred coins of the same diameter to represent fish. Nickels work well.

One small container per team to represent their ship. Paper coffee cups work well. Number the containers sequentially with prominent, easily visible numerals (1, 2, 3, and so on);

Ten slips of paper per team. Index cards work well.

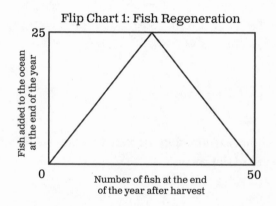

Flip Chart 1: Fish Regeneration

Fish added to the ocean at the end of the year

25

0

Number of fish at the end of the year after harvest

50

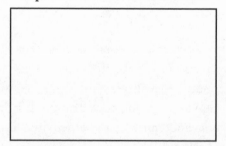

Flip Chart 2: Total Team Harvest

This is blank, except for the title, and it is used after the game to report results.

Two flip-chart sheets showing the information given below. Place the sheets in the room where they can be easily seen.

Setup

Put fifty coins in the ocean. Put the remaining coins in a nearby container that is not accessible to participants. Put ten blank slips of paper in each team's ship. Divide players into roughly equal teams, two to six teams, each with two to seven members. Assign each team a number: 1, 2, 3, and so on. Teams can sit or stand anywhere in the room. They should be far enough from one another that no team overhears another team's strategy. They should also be close enough to the front of the room that they can see the charts and follow your instructions. If time allows, you may also encourage participants to give a name to their ship.

Instructions and Script

Step 1: Create the teams that will play the game. Ask the members of each team to stand near their fellow members. Introduce the exercise by saying, "*Congratulations! Each of you has just become a member of a fishing company. We start with a bountiful ocean.*" (Hold up the container that represents the ocean and shake the coins in it loudly.) "*Your team's goal is to maximize its fish harvest over the course of the game. For this purpose, each team has a state-of-the-art fishing ship.*" Hold up one of the containers you have chosen to represent the ships.

Now slowly read aloud through the following rules and then answer any questions.

Rules of the Game

1. Your team has the role of a fishing company.
2. Your team's goal is to harvest the maximum number of fish for itself over the course of the game.
3. The ocean can support a maximum of fifty fish at one time. The game starts with between twenty-five and fifty fish in the ocean. The precise number is unknown to you.

4. Each round represents one year. You will play for six to ten years. The precise number of rounds is unknown to you until the end of the game.

5. Your team will make one desired harvest request each year. Each request must be some integral number between zero and eight fish.

6. Each round you will write the number on a slip of paper, put the slip in your ship, and bring your ship to me.

7. I will rearrange the ships randomly, then consider each request in sequence.

8. If a request exceeds the number of fish remaining in the ocean, that ship will receive no fish that year. When a team's request can be filled, I'll put the requested number of fish in their ship.

9. Ships are returned to their respective teams.

10. Fish are regenerated according to flip chart 1.

11. The next round begins unless the final round has been completed.

Explain the curve in flip chart 1: *"The curve means that if there are no fish left in the ocean after all orders have been filled, then no new fish will be added to the ocean. But if, for example, there are twenty-five fish left after all requests have been filled, then twenty-five new fish will be added, to reach the ocean's carrying capacity of fifty. If there are thirty-eight fish remaining, twelve will be added."*

Step 2: Explain, *"Now your team should decide on its long-term strategy, and then we will start the first round. Decide on your harvest request for the first year. Write the number on a slip, put the slip in your boat, and send it to me."* Give the teams a few minutes to discuss their long-term strategy and to submit their first fish request.

Step 3: After gathering all the ships, place them on the table in front of you, close your eyes, and rearrange the ships randomly. Open your eyes and straighten out the resulting mass of the ships into a straight line with the ship numbers visible to all participants. You do this mixing because it is important that you fill requests in random order. Ship 1 should not necessarily be the first one to have its orders considered, nor is the first team to hand in its ship guaranteed to have first call on the remaining fish.

Step 4: Pull the paper from the leftmost ship. Do not reveal the size of the request. If there are enough coins in the ocean to fill the request, remove the requested number of coins and put them into the corresponding ship. Then fill the orders from the next ship in the line, if you can, and so on. If one order is larger than the number of fish remaining in the ocean, return that paper to the ship with no coins and go to the next ship. When you have processed all the orders, ask the teams to retrieve their ships.

Step 5: Ask the teams to decide on their next order and send you their ships. While they are doing that, count the number of coins in the ocean and consult the regeneration curve to decide on the number of new fish to add to the ocean. This is quite simple. For a fish population between twenty-five and fifty, you add enough coins to bring the total in the can back up to fifty. Below twenty-five fish, you add a number equal to the number remaining in the ocean after processing all the orders. For example, if there are twelve coins left in the can, you would then add twelve more coins.

Step 6: Collect the ships for round 2, Year 2, process the orders, and continue. If the teams quickly catch all the fish, let them go through one or two more yearly cycles experiencing the consequences of their mistake—no catch. Then stop the game. If you can see that the entire group has adopted a strategy that will keep the fish population sustained around the point of maximum regeneration, you can also stop the game. But with most groups, you will have to go through at least six to eight cycles before participants experience the consequences of their decisions.

Debrief

Typically, one or two teams will pursue an aggressive strategy and place large orders early in the game. That causes the fish population to decline, pulling down the possible harvest for everyone. Sometimes there will be a serious effort to coordinate all the teams' decisions and produce a total harvest that can be sustained over the entire period of the game. But that effort usually fails. Either it is ignored by one or two teams, or it is based on a false estimate of the maximum number of fish that can be harvested annually.

Discuss the regeneration curve (chart 1) with your participants. The regeneration curve shows that twenty-five is the maximum number of fish that can be added to the ocean each year. Therefore twenty-five fish per year is the maximum number that can be harvested sustainably. Over ten years, 250 fish could theoretically be harvested without reducing the fertility of the ocean. Dividing that number by the number of teams provides an excellent approximation of the maximum total fish possible per team. If the total average harvest falls below 25 fish per year, it is the consequence of overharvesting.

Have each team report its total harvest on the flip chart in front of the room. Then lead the participants through a discussion about their experience.

- What happened in this game?
- How was team psychology responsible for this result?
- How were the rules of the game responsible for this result?
- What would have been the maximum possible harvest available to all the teams in this exercise?
- What harvest did teams actually achieve?
- Was there a winner in the game?
- What policies would you have to follow to achieve maximum harvest for all the teams? Why might these policies not be followed?
- Where do you see examples in real life of the behavior we witnessed in this game?

- What policies could be followed in real life to produce a more sustainable use of renewable resources?

Ask participants to think of real-life policies related to commons such as climate and atmosphere.

- *"What is the climate counterpart to maximum sustainable yield?* **The ability of the atmosphere to assimilate GHG.** *What policies could be followed in real life to produce a more sustainable use of this commons?"*
- *"A free rider is a player who attempts to gain the long-term benefits of the group's policies without personally paying the short-term price required to implement those policies. Free riders can derail compromises a group may be trying to negotiate. Where do you see free riders in climate change scenarios?"*
- *"In this game participants experience how quickly and unexpectedly a commons can collapse. How can you effectively communicate the potential for quick, sudden changes in climate?"*
- *"What insights did you gain about how groups can improve discussions when they are attempting to define and solve problems involving complex natural/social systems?"*
- *"What methods are available for overseeing and caring for common resources?"*

In the Harvest game, overfishing produces an immediate and clear gain at first, while the consequences of harvesting more fish than can be regenerated is unclear and delayed for several rounds. How can climate-related policies and actions overcome that time delay and have prompt, positive results?

11.

Hit the Target

Delays between perception and response
can lead to overshooting the goal.

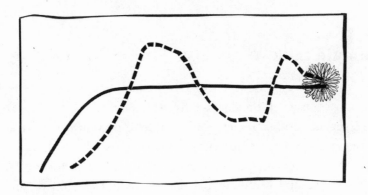

Quotes

The three causes of overshoot are always the same, at any scale from
personal to planetary. First, there is rapid change. Second, there is
some form of limit or barrier. Third, there is a delay or mistake in the
perceptions and the responses that strive to keep the system within its
limits. These three are necessary and sufficient to produce an overshoot.

—Donella Meadows, Jorgen Randers, and Dennis Meadows,
Limits to Growth: The 30-Year Update

It is not enough to aim, you must hit.

—Italian proverb

The climate system contains extremely long delays: It takes time to develop more efficient vehicles and buildings and new carbon-neutral technologies, and still more time to replace existing energy-consuming and energy-producing infrastructure with these new technologies. There are additional delays between emissions reductions and changes in atmospheric greenhouse gas concentrations, between greenhouse gas concentrations and average global temperatures, and between temperatures and harmful impacts such as changes in ice cover, sea level, weather patterns, agricultural productivity, extinction rates, and the incidence of diseases.

—John Sterman, director, MIT System Dynamics Group

Climate Link

Today's weak response to rising GHG emissions is often justified by assuming that when the damage from climate change finally appears to be getting serious, society will be able to prevent it from growing more grave through rapid action. This belief is false and dangerous; it ignores the dynamic consequences produced by sequential delays in perceiving and responding to climate issues. Within the climate system there are many delays: from rising emissions to higher GHG concentrations, from higher concentrations to greater atmospheric heat, from rising heat to growing ecological harm.

Delays in perceiving damage, developing scientific consensus about it, crafting political consensus, and implementing new policies compound the difficulties. Such lags will unfortunately make it impossible to keep climate damage within acceptable limits. When the perception of the destruction finally does become widely shared, it will be too late to prevent problems from growing more serious.

About This Game

We designed this exercise, Hit the Target, to illustrate how even a few delays between perception and response will prevent one from quickly and accurately reaching a goal. It shows how overshoot becomes inevitable when there is a series of delays between action and ultimate consequences. As the delays grow longer, the overshoot increases.

To Run This Exercise

Number of People
This is a demonstration game. You will involve three members of your audience in the demonstration, while any number of people can act as observers.

Time
Twenty to thirty minutes.

Space
You need only room enough in front of the audience to conduct an exercise that involves three participants and a large whiteboard, flip chart, chalkboard, or other writing surface.

Equipment
Large writing surface;
Marker large enough for everyone to see;
Two blindfolds that are comfortable to wear for periods of up to fifteen minutes;
Three name badges: Administrator, Politician, Scientist.

Setup
Assemble the equipment and prepare the name badges.

Instructions and Script

Step 1: Draw the largest possible circle in the middle of the writing surface. The circle should be at least 2 feet (60 cm) in diameter. Larger is better. Put a point in the middle of the circle. Inside the circle write the word "Troubles." Outside the circle write the word "Chaos."

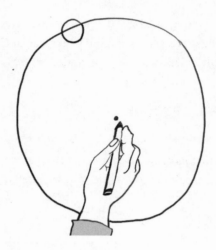

Step 2: Announce, *"I need a volunteer from the audience. The volunteer should draw with his or her right hand and be comfortable wearing a blindfold for a few minutes."* If no one comes forward quickly to help you, simply ask someone in the front row who meets your two criteria to join you in front of the audience. Give this first volunteer a marker. Tell the audience this person represents the Administrator. Hand him or her the appropriate name badge.

Say, *"There will be three trials. At the beginning of each trial you will hold this pen in your right hand in contact with the surface on the dot in the middle of the large circle. Then I will indicate a target point on the circle. When I say 'Go,' your goal is to move the tip of your marker from the middle point to the target as quickly as you can, stopping only when the point of your marker is precisely in the center of the target. I will tell you when to stop. Keep your marker in contact with the surface during each trial, so that everyone can see the path you took from the middle point to the target."*

Explain that this exercise is a metaphor for climate change: *"The starting point in the center of the circle is the GHG level that exists today. The target that I will draw each time at some point on the outer circle is a level of GHG that would give sustainability. The time elapsing as the Administrator's pen passes from the center of the circle to reach the target is the time society experiences troubles before it achieves climate sustainability. Of course if the Administrator over-shoots the target, we will experience chaos.*

"Obviously, we want to be finished with the troubles as quickly as possible. Thus, I ask you to move from start to target as fast as you can. I will ask a member of the audience to time you during each of the three trials, and the chart will indicate your performance."

Step 3: Identify someone in the audience who has a watch with a second hand and ask that person to assist you by measuring the length of time required by the Administrator to reach the goal in each of the three trials. Designate your first target by drawing a small ½-inch (1.25 cm) circle somewhere on the line that forms the large circle. Then say, *"Go!"* In this first trial the Administrator should be able to move the pen quickly and directly from the starting point to the center of the target. When the pen is resting completely within the small target circle, say *"Stop!"* Ask the timer to announce the length of time it took. Label the line 1, and write down next to the line in the middle of the circle the time that was required to move from the center to the target.

Step 4: Ask a second member of the audience to come up, someone who is right-handed and who will eventually be comfortable wearing a blindfold. This person represents the Politician. Have the new volunteer put on the corresponding name tag and face the drawing surface. Put a blindfold on the Administrator, the person who is holding the marker. Move the Administrator's right hand so that the marker is again on the dot in the center of the circle. Place the Administrator's left hand on the right index finger of the new person, who is standing to the Administrator's left and who has his or her back to the audience while facing the surface that displays the circle. The second person, the Politician, can see the circle but is not permitted to speak. The Politician can communicate with the Administrator only by moving his or her right index finger in the direction the pen should move.

Say, *"Again we have the same goal: hit the sustainable target, which I'll draw on a different part of the circle. After I say 'Go!' the person representing the Administrator will try to move the marker as quickly and accurately as possible from its starting point in the center of the circle to a new target that I will indicate on the outer circle.*

"However, the Administrator is now blindfolded and cannot see, so the Administrator has to be guided by the Politician—the Administrator will have to find the target using only the information from movements of the Politician's finger. The Politician can see the circle, of course, but is not permitted to speak and cannot touch the circle. Our timer in the audience will again measure how long it takes."

Draw a new target on a different part of the large circle quite far from the first target. Make sure your timer is ready with a watch. Then say, *"Ready? Go!"*

Make sure the person drawing the line retains contact with the surface so that you and the audience will later be able to see the line that marks the progress from the starting point to the target. Success in reaching sustainability should still be possible, though it will take longer. The line will have more wiggles and perhaps overshoot the target, moving into the area labeled "Chaos" briefly before coming to a rest in the center of the target. Be sure that you tell the timer to stop

only when the pen is fully within the small target circle, not merely close to the target. Label the new line 2, and write down next to the second line the time required to complete the second task.

Step 5: Now ask a third member of the audience to come up. This person will represent the Scientist. Hand over the remaining name badge.

Leave the blindfold on the Administrator, who continues to hold the marker. Put a new blindfold on the Politician, the second person. Guide the left hand of the second person onto the right index finger of the third person. Guide the pen in the right hand of the Administrator back into the central spot inside the large circle. The Scientist has his or her eyes uncovered while the first and second volunteers remain blindfolded and must rely on their touch. The goal remains the same. After you designate a new target quite far from either of the first two, say *"Go!"* The Administrator should move the marker as quickly as possible from the starting point to the center of the target. But this time guidance will come from farther away and involve two other people, which means longer delays. The result will most likely be a line that is much more crooked, and it will likely take the Administrator

much longer to complete the task. More time will probably be spent in the area of chaos. After the target has been reached, label the line 3 and mark down the time required to create it.

Step 6: Now you can tell the two volunteers to remove their blindfolds. Collect the name tags, thank all three of them for their wonderful efforts, and ask them to return to their seats.

Debrief

- *"What general trends did you notice when looking at the three lines?"*
- *"What caused the last effort to be so much slower and less precise than the first one?"*
- *"Where do you see delays in the chain between emissions and ecological damage in the climate change system?"*
- *"What does this exercise tell us about the process of controlling greenhouse gas emissions?"*
- *"As the processes of developing scientific and political consensus about appropriate measures grow longer, should we expect that we will see more or less damage from climate change?"*

12.

Living Loops

It's easier to reach your goals by building
a system that achieves them for you.

Quotes

A positive feedback loop occurs when a small change leads to an even
larger change of the same type. For example, a modest amount of
warming melts ice in northern climates. But the bare ground absorbs
three times as much heat as ground covered by snow or ice, so the
change amplifies the original warming. Even more ice melts, more
heat is absorbed, and the spiral grows.

—Nicholas Kristof, *New York Times* columnist

We must always be on the lookout for perverse dynamic processes which carry even good things to excess. It is precisely these excesses which become the most evil things. . . . The devil, after all, is a fallen angel.

—Kenneth Boulding, economist

You never change things by fighting the existing reality. To change something, build a new model that makes the existing model obsolete.

—R. Buckminster Fuller, mathematician

Climate Link

What causes the earth's climate to change? To answer this question, one must understand feedback. Here the term "feedback" does not mean giving praise or criticism, as in "My teacher gave me feedback on my homework." As a systems term, feedback refers to the circular processes of cause and effect that cause different patterns of behavior: stability, growth, or decay.

Feedback loops—particularly positive, reinforcing feedback loops— are significant contributors toward warming global temperatures. Here is one example of a positive feedback loop from the Arctic: As average temperatures rise, warming melts ice in the northern climates. More ice melts in the summer and less ice forms in the winter. When the sun reaches the earth, some is reflected back to space and some is absorbed. With less ice and more open water, less reflection happens, and so more heat from the sun is absorbed by the open water. As a result, the water temperature rises further. This causes more ice to melt, which results in more open water, which absorbs more of the sun's energy, which causes the water temperature to rise further, which causes more ice to melt, and so on.

About This Game

The Living Loops exercise helps participants understand the struc- ture and behaviors inherent in simple positive and negative feedback loops.[10] The exercise also illustrates how individuals and groups can use feedback loops to achieve desired goals.

Living Loops can be used as a quick and simple way to boost understanding of how systems function, demonstrating that the behavior of a system is produced by the interrelationships within that system and that a change in one element in the system can alter the behavior of that system. Participants are encouraged to hypothesize about the impact of a simple change in feedback (from positive to negative, or vice versa) or in the type of loop (open or closed) and then to test their hypotheses. Often participants walk away with the realization that one part or one person can make a difference in even the most complex systems.

You may use Living Loops to:

- Illustrate through participants' own physical movements the behavior of balancing and reinforcing feedback loops;
- Link physical experience with intellectual analysis of behavior in closed chains of cause and effect;
- Develop a more intuitive understanding of the basic dynamics in simple feedback systems.

To Run This Exercise

Number of People
This is a demonstration game. It will accommodate five to twelve as players and any number of observers.

Time
About thirty minutes.

Space
A room free of obstacles in which a group of five to twelve people can stand shoulder to shoulder in a single line or in a circle holding hands. There should be enough room for observers to watch the action and hear you, the facilitator. Move chairs and tables to the sides of the room, if necessary, to give people enough space to play and observe.

Equipment

One ball or other colorful, easily held object;

One index card per player; observers need none;

One piece of string per player—long enough to fit generously over the head, like a necklace;

Stapler or roll of clear tape.

Setup

Attach both ends of each string to an index card with a stapler or clear tape. Using a marker, label each card with a big + on one side and a big – on the other side.

Things to Consider

It is always important to attend to participants' physical and psychological comfort. This exercise involves gentle movement: players should be able to move their arms and hands easily from below the waist to high over the head. Before you ask for volunteers, point out that the activity will involve reaching and some bending. An observation role is always available to those who do not wish to participate. Observers stand or sit outside the circle, watch players' movements, and ensure that players move in accordance with the rules of the game. You can also adapt the exercise in advance for physically challenged players. For example, if a player uses a wheelchair, you could easily conduct the exercise with everyone sitting down.

Instructions and Script

The instructions for Living Loops are detail intensive. We encourage you to read them carefully several times before conducting the activity.

Note: In this exercise it is particularly helpful to debrief as you go along.

Step 1: First you will create and demonstrate an open loop. Invite five to twelve volunteers to come up and serve as players. Have players stand in a line, shoulder to shoulder, facing the observers.

Take one of the index cards with a string attached, put it over your head like a necklace, and flip the card so that the + sign faces out on the front of your torso. Ask the players to do the same.

Place the ball or other visible object in the left hand of the person standing at the far right end of the line when viewed from the audience. The ball will be held and observed but not passed in this game. Walk to the other end of the line and get in line yourself, facing the audience. For the audience, you will be on the far left.

Explain that every player's left hand is going to be "active." Ask players to clench their left hands into fists and hold them out at waist height. Do that yourself.

Explain that players' right hands are "passive." Ask everyone to rest his or her right hand lightly on the fist of the person to his or her right. Extend your own right hand out from your waist, open and palm down.

Explain that the signs on the cards indicate the nature of the response by each player's left hand to the incoming signal. If a person is wearing a + sign, his or her left hand must move in the same direction (up or down) and distance (number of inches) as his or her right hand, after a one-second delay. For those wearing a – sign, their left hands must move in the opposite direction and same distance as their right hands, after a one-second delay.

Demonstrate. Point out that everyone is wearing a + sign, so all their left hands must move in the same direction as their right. Send a practice "signal" or pulse to the person on your left by raising your right hand 2 inches (5 cm) above waist level and then, one second later, raising your left fist 2 inches. Point out that the person on your left, having felt his or her right hand rise 2 inches, must now move his or her left fist up by the same distance after one second. Then each person along the line will respond successively until the last person, who is holding the ball, finally raises it 2 inches and stops.

Invite everyone in line to lower their hands and shake out their arms to relax. Ask if they have any questions.

Now explain that you're going to lower your right hand by 2 inches. Lower your right hand, then after one second your left hand, and ask the rest of the players to follow suit in sequence. Ask observers to note the positions of everyone's hands and the ball at the end of the line over the time period of the exercise. The ball should eventually drop 2 inches from its starting position and then stop moving.

Now change the sign of one person in the line by flipping his or her label card from + to –.

Repeat the definition of "active" and "passive" hands and of + and – links. Announce that you are going to raise your left hand just once by 2 inches. Ask what the ball will do. Give participants and observers time to make their predictions. Then move your left hand up by 2 inches. Check to make sure that each successive person, starting with the player on your left, moves his or her fist in the proper way down the line. Do not let people anticipate the motion; they should move their left fist only one second after they have actually felt their right hand move. Eventually the ball should fall 2 inches and then stop.

Step 2: Next you will close the loop and demonstrate its behavior.

When the signal comes to the end of the line, observe: *"In order to move the ball all the way to the floor, we would need to make continual inputs from the other end of the line. Now let's see what happens if we close the loop."*

Make sure all signs are showing +. Now ask the participants to form a circle. The player holding the ball will be standing next to you on your right. Place your right hand on top of that person's left hand, which is holding the ball. Point out explicitly that you will make only one independent move and thereafter simply obey the rule of motion dictated by your sign.

Raise your left hand 2 inches and watch that impulse move around the circle until it reaches you, causing your right hand to raise 2 inches. Then obey the rule of your sign, a +. You will naturally move your left hand 2 inches higher. Let the signal travel all the way around the circle several times, each time carrying the ball 2 inches higher. Eventually, a player will reach the limit of his or her ability to reach higher, and the signal will have no option but to stop.

Ask the players to release their hands. Invite them to describe what happened.

Ask, *"Why was the behavior of the ball so much different this time?"* The answer, of course, is because you closed the loop for the first time, creating feedback. Since all the links were +, the group created a reinforcing loop.

Ask, *"When have you felt as if you were in a reinforcing structure in an everyday life situation?"*

Because the loop is closed, the structure of the system has taken over. It does not require continual input from the outside.

Demonstrate this idea by running the exercise again. Announce that the initial motion will be 2 inches down. Give people an opportunity to predict what will happen. Most of them will correctly guess that the ball will this time move progressively down until someone's hands reach the floor. Then start operating the loop by lowering your left fist. The ball will move downward by 2 inches each cycle until someone's left hand encounters the floor.

Point out that all reinforcing structures have limits that determine how far they can grow in one direction. People will physically experience the idea that limits are inherent in reinforcing loops.

Step 3: Finally, you will change the sign of players to contrast the stabilizing influence of a negative loop with the amplifying influence of a positive loop.

Explain that the group will run another experiment. Say, *"This time we will introduce a single negative link into our system. Most of you will still be representing a positive link and so must still move your left fist in the same direction and distance as your right hand. One of you will represent a negative link. That person will move his or her left fist in the opposite direction but same distance as his or her right hand."* Point to one person and tell that player to switch his or her sign from + to − (by taking off the card, reversing it, and putting it back on). Have the group remain in a circle.

Announce that the input will be 2 inches down. Ask players to figure out how their own fists will move as the signal travels around the circle. Let players think about this silently for a few seconds. Then invite a few volunteers to demonstrate what they think their fists will do over the time period of the exercise.

Ask everyone to arrange their hands as before. Move your left hand down 2 inches. Watch to make sure that the − link behaves correctly and that the others are also correctly following the movement. This switch from + to − in one link causes people to move their hands in the same direction as the initial impulse until the signal reaches the − link. There the direction is reversed. Those after the − line will move their fists upward until the signal has passed all around the circle and again encounters the − link. Thereafter, for a cycle the direction is downward and then up afterward, then down and then up, as the signal passes around the circle and through the − link. Stop the group after several rounds as soon as everyone sees that the loop is oscillating indefinitely.

Make the following observation: *"Simply by changing one sign, we changed this from a reinforcing loop, which amplifies the initial input, to a balancing loop, which tries to offset or correct it. If we compare the open loop to the closed loops, we see that the open loop needed continual input to sustain change whereas the closed loops, whether balancing or reinforcing, progressed with no input after the initial impulse."*

Invite the group to drop their hands and shake their arms to relax. Ask the following questions:

- *"What happened?*
- *"How long do you think this system behavior could continue?"* (The answer is forever.)
- *"What was the difference between the open and closed loops? And between the balancing and reinforcing loops?"*
- *"What did it feel like to be in this balancing loop, compared to how it felt being in the reinforcing loop?"* Explain that the normal behavior of a balancing loop is oscillation; that is, the "motion" of the system continuously moves back and forth around a fixed point, just as people's hands were oscillating during the exercise.
- *"When have you felt as if you were in a balancing structure in an everyday-life situation?"* If participants are having trouble coming up with examples, offer the example of hunger and eating, savings accounts and spending, exercise and stress. Then ask players to try to generate more of their own examples.

Debrief

This exercise gives participants direct experience of the behavior that may be expected from the many potential tipping points in the climate system. They all involve positive feedback loops, reinforcing loops that can sustain their own behavior without any further input from human action. There are many examples of climate tipping points that are of possible concern. Below are three examples.[11]

Ask: *"In the game, a small initial effort was sufficient to take the ball all the way to the ground when its influence was reinforced by a positive loop. What are some of the positive, reinforcing loops that might be created to assist in reducing GHG emissions?"*

One example would be to create better bicycle paths within a community. As the ease of commuting on a bicycle increases, more people will switch to bicycles from their cars. As the bicycle traffic goes up, political support for bicycle paths will increase. That will lead to creating even more paths.

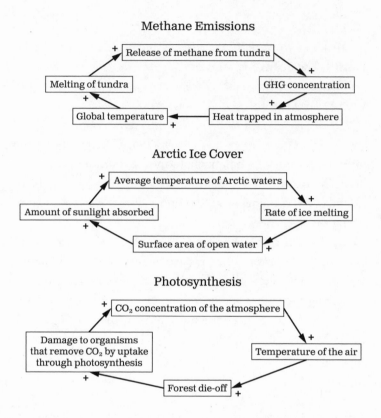

Ask: *"In the game, the negative loop functioned to counteract an initial change, and keep the ball in its initial position. What are some of the negative, balancing loops that might be created to hold climate parameters within habitable limits?"* One example would be to monitor the air temperature within a community. As average temperatures increase, one could invest more in changing the color of surfaces to lighter shades that reflect more light, letting less sunlight be converted to heat and thereby reducing temperatures below the level they might otherwise have attained. Another example would be to tax carbon emissions, raising the tax as GHG concentrations increase. Raising the price of energy-intensive goods will reduce demand for them, lowering their manufacture and thus reducing emissions.

Paying attention to balancing dynamics will help you anticipate resistance to your efforts. As you increase efforts to change some aspect of the current system, those who benefit from it will increase their efforts to resist your change. In what ways could climate change efforts cause reactions in the wider system that could reduce or weaken your desired results? Should you avoid this resistance? If so, how? Would your understanding of the possible resistance to one action lead you to choose a different action?

13.

Paper Fold

With exponential growth, small growth rates can
quickly lead to extremely large numbers.

Quotes

People at present think that five sons are not too many, and each
son has five sons also, and before the death of the grandfather there
are already twenty-five descendants. Therefore people are more and
wealth is less; they work hard and receive little.

—Han Fei, philosopher

Bacteria multiply geometrically: one becomes two, two become four,
four become eight, and so on. In this way it can be shown that in a
single day, one cell of *E. coli* could produce a super-colony equal in
size and weight to the entire planet Earth.

—Michael Crichton, *The Andromeda Strain*

The most powerful force in the universe is compound interest.

—Albert Einstein, theoretical physicist

Climate Link

One dilemma for climate change activists is talking about policies that seek apparently insignificant reductions in activities that people value, such as energy use. The problem is that many of the activities that lead to rising GHG emissions exhibit exponential growth. And even small exponential growth rates quickly lead to extremely large numbers. The exercise helps people understand the dynamics of exponential growth. And it can be used to illustrate the concept of doubling time. Doubling time is the length of time required for a growing entity to double in size. It is estimated through dividing the number 72 by the rate of growth in percent.

Citing a growing entity's doubling time, rather than its annual growth rate, helps most people have a better understanding of its potential for future expansion. For example, GHG emissions have recently increased about 3 percent per year. That growth rate might seem innocuous, yet that rate corresponds to a doubling time of about twenty-four years. You may state that something is growing at 3 percent per year or you can point out that it will increase more than sixteen times over a century. The latter seems much more significant.

About This Game

We adapted Paper Fold from a classic brain teaser, often presented as a simple trick. We use it to illustrate dramatically the power of exponential growth. The main challenge in using this game is to get participants to accept that its explosive results actually convey useful information about climate systems. This is difficult, since rapid growth in Paper Fold occurs about five million times faster than in the climate system.

To Run This Exercise

Number of People

This is a mass game. It can be observed by any number.

Time

Five to fifteen minutes.

Space

This exercise is typically carried out while participants are seated.

Equipment

One bed sheet or large tablecloth for use by the facilitator. Do not use a large piece of paper, such as a page from a flip chart, since it is too thin to give an observable thickness after only four folds.

Setup

Have the cloth you will be folding at hand.

Instructions and Script

Step 1: Provide some justification for the exercise you are about to conduct. Say, *"We have been talking about an issue that involves behavior over the long term. Let me show you now an exercise that illustrates some important points about long-term behavior."*

Pick up the item you will fold and hold it open, with no folds. *"Here I have a bed sheet [or tablecloth]. It is very thin."* Show them the edge of the item, so they can see how thin it is.

Step 2: *"Now I fold the sheet in half once. A second time. A third time. And a fourth time."* Actually perform the four foldings as you talk about the process.

"Each fold doubles the previous thickness. After four folds, the sheet is about half an inch [or 1.25 cm] thick." Hold it up edgewise, so that participants can see the thickness. Hold it up loosely, if necessary,

in order to avoid pinching it down to a thinner cross section. The numbers that appear in the following script depend on the fact that the item is plausibly half an inch thick after four folds.

"Of course, you could not physically fold this sheet in half twenty-nine more times, for a total of thirty-three folds. But imagine that you could. How thick would it be then? After four folds it is half an inch. How thick will it be after twenty-nine more folds?"

Step 3: Invite answers from the group. *"All those who think the sheet will reach from the floor to below my waist, raise your hand."* Pause; look around the audience. You should see a few hands if people are indicating their honest opinions. *"All those who think the sheet will reach from the floor to below the ceiling, raise your hand."* Pause again and look around the audience. Someone might say something like, "To the moon." If you hear this claim, state emphatically, *"No, not to the moon. Not even close!"*

Then tell them the answer. *"If four doublings take the sheet to half an inch thick, twenty-nine more doublings would make the sheet more than 3,400 miles [or 5,000 km] thick. That is approximately the distance from the city of Boston in the United States to Frankfurt, Germany."*

Debrief

Most participants consider the correct answer preposterous and assume there is a trick behind it. Therefore, in debriefing the exercise, you may want to try first demonstrating the math behind the answer. Use slides or a white board to show the dramatic outcome of starting with 1 and doubling it 29 times: 1, 2, 4, 8, 16, . . 536,870,912. Doubling something twenty-nine times increases it by a factor of about 540 million. After four folds, the sheet is about half an inch thick. Doubling it twenty-nine more times would produce a thickness of 216 million inches (about 549 million cm). A mile is about 63,400 inches (161,000 cm), so the folded sheet would be a little over 3,400 miles thick.

You can quit at this point, having demonstrated that the process of doubling quickly produces unexpectedly enormous numbers. Or you can spend some time countering the tendency to believe that any process that involves doubling, a 100 percent growth rate per fold, cannot be relevant to a process that involves "only" a few percent growth per year. Point out that an expansion of 4 percent per year will increase something by more than fifty times over a century and twenty-five hundred over two centuries.

We often ask people to draw what is called a "behavior over time" graph for the increasing thickness of the sheet, assuming that they could accomplish one fold every second for thirty-three seconds, and assuming that the sheet initially has a thickness of half an inch. Point out that it seems like nothing much is happening during the first 80 percent of the process. That is our problem with climate change. The small annual increments do not seem significant. But they imply massive changes eventually. Ask where else members of your audience have seen this sort of behavior. Population growth and growth of energy use are both examples.

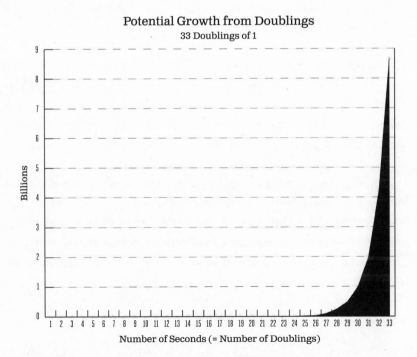

Potential Growth from Doublings
33 Doublings of 1

Depending on the time available, you may want to elaborate on the issue of population growth, since it is dramatic and piques many people's interest. You can say, *"We chose to illustrate thirty-three doublings in this activity for a reason. Today's global population is almost thirty-three doublings from the first person on Earth. More than seven billion people currently live on the planet."*

If you wish to continue, tell them the riddle of the water lily. *"A traditional French riddle also illustrates the surprising nature of exponential growth. Suppose a water lily is growing on a pond in your backyard. The lily plant doubles in size each day. You are told that if the lily is allowed to grow unchecked, it would completely cover the pond in thirty days, choking out all other forms of life in the water. For a long time, the plant seems small, so you decide not to worry about cutting it back until it covers half the pond. How much time will you have to avert disaster, once the lily crosses your threshold for action? The answer is one day. The water lily will cover half the pond on the twenty-ninth day; on the thirtieth day, it doubles again and covers the entire pond. If you wait to act until the pond is half covered, you have only twenty-four hours before the lily chokes out the life in your pond."*

The behavior in all of these instances seems counterintuitive. We generally expect things to follow linear patterns of growth. For example, the height of a pile of paper grows linearly when new sheets are added to the top of the pile at a constant rate. With linear growth, the amount of physical change initially is the same as the amount of physical change later in the process. But reinforcing processes produce nonlinear growth, and they can quickly convert a small initial change into an enormous change. In folding the sheet, no significant change is noticeable for many doublings. Then, although the underlying growth process hasn't changed at all, an explosion seems to occur. The thirty-fourth doubling would actually add another 3,400 miles to the sheet's thickness.

14.

Paper Tear

One-way communication is much
less effective than interaction.

Quotes

There are still, thankfully, at least a handful of prominent reporters
who understand climate change from soup to nuts. Their work, and
quite frankly, their jobs, becomes more significant as widespread,
impoverished mass communication dramatically and rapidly under-
mines climate policy of any kind at home and abroad.

—Eric Roston, "Climate Change: Melting Ice Makes
Slippery Slope," *Grist*, February 2010

Information is the key to transformation. That does not necessarily mean *more* information. . . . It means *relevant, compelling, select, powerful, timely, accurate* information flowing in new ways to new recipients, carrying new content, suggesting new rules and goals. . . . When its information flows are changed, any system will behave differently.

—Donella Meadows, Jorgen Randers, and Dennis Meadows,
Limits to Growth: The 30-Year Update

The single biggest problem in communication is the illusion that it has taken place.

—George Bernard Shaw, playwright

Climate Link

The demand for long-term action to avoid serious climate disturbance on a worldwide scale must be the most complex and urgent challenge society has ever faced.

Climate change poses a particularly challenging set of concepts to convey. Not only are its effects invisible for many people, but there are significant time delays between action and consequence. People inevitably draw on their experiences of weather in their efforts to understand climate, yet weather (a short-term phenomenon) and climate (a very long-term average of weather patterns) are fundamentally different. The causes of climate change extend across disciplinary boundaries, so parties to the discussion often do not understand one another's vocabulary, and a well-funded set of organizations is exploiting every disagreement and controversy to confuse the issue and block change. Understanding the threat of climate change requires scientific knowledge outside the cognizance of most citizens, and it poses moral challenges outside their experience, even their concern.

These obstacles must be overcome if global society is to deal proactively with climate change. Never before has effective communication and education been so necessary.

About This Game

This exercise illustrates that efforts to communicate even simple ideas can fail. It helps people understand the reasons for poor communication and practice some of the skills around effective communication. When stakeholders come together in a group, they frequently discover they have different perceptions of what is going on. This is especially true when the group hopes to understand and intervene in a complex system. As divergent perspectives become evident, those trying to communicate tend to increase the frequency and the volume with which they express their thoughts. Instead, they should try to empathize with listeners and discern what they understand and anticipate possible sources of misunderstanding.

It is generally assumed that an eloquent speaker will leave his or her listeners with the same images in their minds. This simple exercise shows quickly how unfounded this assumption really is, even when listeners share speakers' goals and have a strong incentive to understand their meaning. How much *more* challenging is communication, then, when the opposite is the case?

This game can heighten listening and communication skills and increase awareness of the multiple interpretations of the same message. It is useful for people working in an area as complex as climate change to reflect on how others hear their words and call to action, both with and without two-way communication and feedback.

This exercise is useful at the start of a meeting to remind participants that communication requires constant care and the involvement of everyone.

To Run This Exercise

Number of People

This is a mass game. It requires a comparison of people's final result, so it does not work very well with just one or two players. A minimum of five players is suggested. The maximum number is unlimited.

Time

Ten to fifteen minutes.

Space

No special space is required. This exercise is typically carried out with an audience of people who remain in their chairs.

Equipment

One sheet of standard letter-sized paper for each person, preferably recycled or scrap. As long as the sheets are all the same size, used paper from the copy machine, whether colored or printed on one side, works well for this exercise.

Setup

Pass the stack of paper around the room and ask each participant to take a sheet. Keep a sheet for yourself. Or if you are in a hurry, distribute the paper ahead of time, putting a piece either on or underneath each seat. When people come in, you can tell them the paper is for use later.

Instructions and Script

Step 1: Ask everyone to pick up the piece of paper. Ensure that all participants are sitting someplace in the room where they can see and hear you.

Step 2: Explain the exercise. *"This piece of paper is a metaphor for our policy options vis-à-vis climate change. I have an extremely important message to give you about a new climate policy that we all need to follow. It is a critical message, and I ask you please to listen carefully and not interrupt me or talk while I share my important message. Do not ask questions! Just do precisely as I ask you to do. Tearing the paper is a metaphor for the policy steps. Our goal is for everyone to produce identical patterns with their pieces of paper."*

Step 3: Hold up your paper for everyone to see. *"Fold your paper in half and tear off the bottom right corner of the paper."* Do that to your paper and pause for a moment to allow the group to do this. Sometimes a member of the audience will ask for more precise instructions. Simply say, *"Please don't ask questions. Just do as I say."* Continue: *"Fold the paper in half again and tear off the upper right-hand corner."* Do that to your paper and pause briefly again. *"Fold the paper in half again and tear off the lower left-hand corner."* Do that to your paper and pause briefly again. *"Okay, you are all intelligent people, eager to follow instructions. Let us see how you did."* Unfold your paper and hold it up. *"Please unfold your paper and hold it up for the group to see."* Pause for an extended period until everyone has had a chance to look around. Of course the precise shape of the torn holes will differ. But that is not important. The key question is whether the pattern of holes is the same from one paper to the next. Normally, it is not. There will be a variety of different patterns. Some will be identical to yours; most will not.

Step 4: Ask participants for their reflections on what happened and what set of behaviors on both your and their parts produced this result. Once they have shared their opinions about what went wrong, ask participants for their advice on how to have a better outcome that more closely reflects the goal for everyone to have the same result at the end of the exercise.

Debrief

Most often each participant creates one of four or five different shapes out of their paper. Participants are likely to be surprised by the different interpretations of the same, simple message and set of instructions.

"When intelligent people fail, the problem is normally in the structure of the process. What was there about this communication process that gave us so little success?"

Give the participants a chance to reflect on this question and respond. The key failures in the communication process were:

- One-way communication. You did not permit questions.
- Ambiguous vocabulary. "Fold in half" and "upper right-hand corner" and other phrases can be interpreted differently. For example, upper right for the speaker, who is facing the audience, is upper left for people who are observing. Is the fold facing up or down when you tear a corner from it? The instructions left out many specifics.
- No common understanding of the final goal. You did not show them the final result you wanted; you focused only on the process.

Make sure that each of these failures is identified, preferably by the participants. Then you can ask questions:

- *"Is it important that people share a vision of the causes and consequences of climate change? Why?"*
- *"Is it important for people to have the same understanding about policy options? Why?"*
- *"Where do these failures also exist in discussions about climate change?"*
- *"How could we make the communication more effective?"*

End with drawing out some lessons that participants can keep in mind when they are communicating about their climate change work. What are some opportunities they see to adopt these lessons and practice these skills in the near future?

15.

Pens

Sustainability depends more on
culture than on technology.

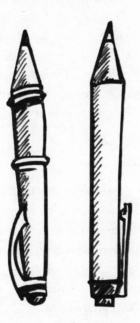

Quote

It's not technology that limits us. We're the limitation. Our technology
is an expression of our intelligence and creativity, so the limitations of
our technology are a reflection of our own limitations.

—Christian Cantrell, engineering manager

We cannot hope to create a sustainable culture with any but sustainable souls.

—Derrick Jensen, author

A bad carpenter blames his tools.

—English proverb

Climate Link

Many people concerned about climate change promote solutions involving changes in technology, such as shifting away from fossil fuels and toward various non-carbon-based energy sources, like wind, solar, and wave power. But even if society could be very successful in implementing new technologies, it would still allow drastic climate change if it tolerates population growth, pursues economic growth, and uses per capita GNP as an indicator of success.

About This Game

This exercise was first created for use in Japan, where Dennis had little time for communicating. And he faced great cultural barriers when trying to communicate with several thousand Japanese corporate and government leaders who had assembled for a several-day conference on technology policy. The group had assembled for a meeting called "Technology for Sustainable Development," a name that revealed an important misconception. Technology is not the main tool for attaining sustainable development. It mainly serves the values and goals of those who develop and use it. If their values and goals promote growth, the technology they develop will produce growth, not sustainable development. Real solutions to climate change and other symptoms of physical growth on a finite planet will first of all require changes in the values and norms of society. This exercise helps to make the point that sustainability is inherently a matter of social norms, cultural practices, and psychological attitudes, not technology.

To Run This Exercise

Number of People

This is a mass game. Any number can observe.

Time

Five to ten minutes.

Space

No special space is required. This exercise is typically carried out while participants are seated.

Equipment

You will need two pens, which you will show the audience: a noticeably expensive pen and a cheap pen. This exercise works well if you use one widely recognized premium-brand, refillable pen and one "green" pen made of cardboard and wood. Alternatively, instead of real pens you may use slides showing photos of the pens, or use two pens of any kind and ask the audience to imagine that you are holding one cheap and one expensive pen.

Setup

Have the pens in your pocket or the slides ready to project.

Instructions and Script

The following instructions and script are written as if you actually use in the exercise an expensive pen made by Montblanc and a cheap pen made from cardboard.

Step 1: Say, *"We are meeting in part because we all share a concern about sustainable development. I am now going to conduct a simple exercise to see if we all agree on what the term 'sustainable' actually means. I have two pens."* Hold them up so that everyone in the audience can see the two pens.

"The first pen is made by Montblanc. It is made from expensive metal and precious resin. It is used for writing. It cost about 400 US dollars. The second pen is made from wood, plastic, and recycled cardboard. It is used for writing. It cost about one US dollar. Which pen is more sustainable?"

Here you need to get every member of your audience to develop a personal answer to your question and actually make a choice. You can make it easy for them to do that with the following instructions. *"Refer to the Montblanc pen as number 1 and the cardboard pen as number 2. Decide which pen you think is more sustainable and show either one or two fingers, silently, to the person sitting next to you."* Pause long enough for them to ponder the issue and indicate their decision to their neighbor. We normally ask for a silent communication with each participant's neighbor because in some cultures people are reluctant to raise their hands in a situation where they may turn out to be "wrong."

Step 2: Continue, *"Now I will provide you with more information. The Montblanc pen is never taken out of my home office, so I will use it the rest of my life and then pass it on to a friend as a tool for him to use. In contrast, I lose a cardboard pen almost every time I take one out of the house. Thus, I must buy many dozens of them each year. When the Montblanc pen's ink cartridge becomes empty, I buy a refill. When a cardboard pen's ink cartridge becomes empty, I throw the pen away and buy a new one. It costs more to buy the replacement cartridge than to buy a complete new pen. Now I will ask you again: which pen do you think is more sustainable? To indicate your decision show one or two fingers to your neighbor, showing one finger for number 1, the Montblanc pen, and two fingers for number 2, the cardboard pen."*

Pause and then say, *"Note that many people changed their votes."* You may be uncertain about this. But some participants probably changed their vote, and no one will challenge this statement.

"However, the new information I gave you did not describe the pen's physical technology; it described my relationship, habits, and

attitudes toward the pens. This exercise demonstrates that you actually believe the property of sustainability is not mainly inherent in the physical technology of the tool. Instead, sustainability is inherent in a person's relationship to the tool. Achieving sustainability may perhaps be easier if we can adopt new technologies, but we could reduce GHG emissions with better use of the technologies we already have. Much more important will be the development of new relationships and attitudes governing the use of the technologies we use."

Debrief

Here are some sample questions for debriefing:

- *"What are some technical measures to reduce emissions of greenhouse gases into the atmosphere? Will these be effective without concurrent social changes?"*
- *"What are some social measures to reduce emissions of greenhouse gases into the atmosphere? Which of these is more widely mentioned?"*
- *"What purely technical measures could stabilize emissions while society continues to promote indefinite growth?"*
- *"What could you do to advance changes in the social and cultural factors that promote growth?"*
- *"If you were successful, how would those changes affect greenhouse gas emissions?"*

16.

Space for Living

Thinking outside the box can
produce win-win solutions.

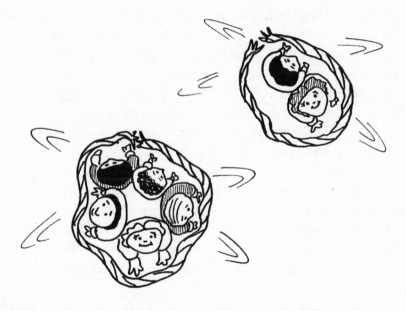

Quotes

The blunt truth about the politics of climate change is that no country
will want to sacrifice its economy in order to meet this challenge, but
all economies know that the only sensible long-term way of developing is to do it on a sustainable basis.

—Tony Blair, former British prime minister

The financial crisis is a result of our living beyond our financial means. The climate crisis is a result of our living beyond our planet's means.

—Yvo de Boer, former executive secretary of the United Nations Framework Convention on Climate Change

Sharing what you have is more important than what you have.

—Albert M. Wells, author

Climate Link

Vanishing natural resources are a commonly shared concern of our climate crisis. There are many examples, from diminishing forest carbon sinks, to falling water tables, to declining habitable spaces in islands and coastal nations, to shrinking Arctic ice fields for polar bears. Humans and other species face mounting challenges related to sharing essential but diminishing resources.

Many well-intentioned efforts to adapt to declining resources founder on three deeply ingrained ways of thinking—conceptual habits that form the way we approach complex problems. First, people tend to avoid thinking and talking about better sharing of resources. A common belief is, "If he gets more, then I'll have less." Second, many people prefer to ignore a problem until it is widely perceived. They think, "When I see it, I'll do something about it." Third, people often rely on tried-and-true methods. They believe, "If it worked before, it will work again."

This exercise can lead a group through a shared experience in which they confront the consequences of these three conceptual habits.

About This Game

This exercise can help people experience their own and others' reactions in the face of declining resources. It helps illustrate some basic principles governing innovation and opinion change in groups and can provide a metaphor that is relevant to situations that arise when resources become inadequate to support the habitual way of doing business. It can demonstrate the need to be open to new plans or policies even while current ones seem successful.

This game is not recommended for use at the beginning of a workshop. Rather, it is best to wait until participants have gotten to know each other a bit before asking them to put themselves in close physical proximity.

To Run This Exercise

Number of People

This is a participation game for a minimum of ten to fifteen but ideally twenty-five people. With groups larger than thirty, split into subgroups of fifteen to twenty participants.

Time

Fifteen to thirty minutes.

Space

Outdoors is best if the surroundings and climate are not too distracting. If you use a space indoors, you need an open floor area of at least 20 by 20 feet (6 by 6 m), clear of any objects. A bigger space would be even better.

Equipment

In total, allocate about 10 feet (3 m) of rope per participant. The first time you play the game, you will cut the rope into pieces of different length and tie each piece into a loop. In subsequent sessions of the game, with other groups, you can reuse the rope by tying several pieces together to get the correct number of loops of the proper size. Cotton clothesline works well, whereas synthetic ropes may easily come untied. Check your knots!

Setup

Take half the rope and cut it into pieces 4 feet (1 m) long. Tie the ends of each piece of rope together to make a loop large enough to encircle one pair of feet placed together flat on the ground, with no part of the shoes touching the rope or the floor outside the rope. Use a fifth of the rope length to make loops large enough to encircle the feet of two to three participants. Each of these loops will require a piece of rope about 7 feet (2 m) long.

Make one loop big enough for about two-thirds of the participants to stand within it very closely packed. For ten people, for example, you need about 15 feet (4.5 m), for twenty-five people 20 feet (6 m).

Use the remaining rope to make loops big enough to encircle the feet of five participants, about 11 feet (3.5 m) each.

Put all the loops on the ground so that they are at least 1 foot (30 cm) away from each other. Pull each loop into the shape of a circle. The number of loops on the ground should be greater than the number of participants.

Things to Consider

This game requires close physical proximity as well as physical mobility. If you believe that one or two people may feel uncomfortable with this game, consider enlisting them to pick up the ropes and check for compliance with the rules. If you think more than just a few participants will be uncomfortable, do not use this exercise. One adaptation that might work well for some groups is to divide them into male and female subgroups so that there is no close proximity among those of the opposite sex.

Watch to ensure that no one creates a solution to the crowding that puts anyone under physical stress or poses the potential for people to lose their balance and fall over. Space for Living has been run hundreds of times with no problem, but it is always wise to be cautious.

During the discussions and the debrief, it is best not to call on a specific person. Let the participants choose to share their thoughts or not.

Instructions and Script

Step 1: Gather your group near the rope loops. If you have more than one group, you will need at least one supervisor for each subgroup.

Ask all participants to go and stand inside a loop so that both of their feet are on the ground and neither foot touches the rope. Best is for each participant to start with his or her own loop, but this is not essential.

Step 2: Explain the game's rules: *"Imagine that the space you're standing in represents an important resource."* You can choose one that is appropriate: a carbon sink that is important for fixing enough carbon for your survival, habitable land on your island, arable land on Earth, enough rangeland for your cattle, water for your crops, and so on. *"To 'survive' to the end of the game, each of you needs to keep finding space within a loop. It is essential in each round that you find a place where your feet touch the ground only inside a loop without touching the rope. Your feet must not touch the ground outside the loop. Anyone who has not found a space within thirty seconds after I say 'Switch!' will be asked to move to the sidelines."*

Explain how each round will proceed: *"When I see that all of you either have space for living or have moved to the sidelines, I will say, 'Switch!' Then, if possible, you must leave the loop where you are standing and find space for your feet within a different loop. I will wait until everyone has either found a space, feet not touching the rope and not touching the ground outside the rope, or moved out of the game. Then I'll say, 'Switch!' again."*

Step 3: Look around to ensure that everyone is standing inside a loop with no illegal touches—no feet touching the rope or the ground outside the loop. Then say, *"Switch!"* Wait for all participants to move to a different loop and position themselves inside it without any foot touching the ground outside a loop.

Step 4: Repeat the exercise a third time. However, just as you say, *"Switch!"* pick up several smaller loops as soon as they're empty. It helps to have a colleague assist you with this. If someone refuses to vacate the loop you wish to pick up, just untie the knot and remove it from around him or her. There may be brief panic until participants realize that more than one person can stand within the larger loops. Observe how this realization first occurs to an individual and then moves through the group.

You might choose to make the analogy stronger and increase drama by announcing toward the end of the game, when you take away a loop or loops, *"We've just lost another 80,000 acres [32,000 ha] of forest from deforestation"* or *"The East Antarctic Sheet just shed another fifty-seven billion tons of ice,"* depending on the metaphor you have chosen.

Step 5: Continue with several more rounds, each time removing several smaller loops. Whenever you observe that someone is not accommodated within a loop thirty seconds after you have said *"Switch,"* ask him or her to step away from the active zone of the exercise and observe. Reassure all inactive players that their observations will be useful at the end of the game. This helps them stay engaged in the play.

Step 6: When only one or two loops are left, it will be impossible for all the remaining participants to stand completely within the space the loops provide. At this point some participants may start creating human pyramids—for example, by trying to carry colleagues on their shoulders. Don't permit this strategy; it is dangerous. Remind people that everyone's feet must touch the ground and everyone must be self-supporting.

Some people may ask whether their feet must be flat on the ground or whether they can stand on tiptoe or position their feet in some other way and still follow the rules. A good reply is: *"Everything that is not forbidden is permitted."* Eventually, someone will realize that it is therefore legal for participants to sit or lie on the ground outside the circle with just, for example, their heels touching the ground inside the circle. Notice how and with whom this idea originates and whether others in the group promote or resist it.

If a high-status participant, such as a senior manager, initiates the idea, often other group members will quickly support it. The same idea coming initially from another participant will often be ignored. Make a note of this to bring back to the debriefing around power asymmetries and communication.

Step 7: When all remaining group members have successfully managed to have their feet touching the ground inside the last remaining rope circle, the exercise is complete. Then lead a brief round of group applause. Help those who are on the floor get up, and segue into the debrief.

Debrief

In order to get the most learning from this exercise, make sure you have enough time to debrief it. You can ask a variety of questions to prompt insightful discussions on three levels. First, focus on participants' behavior:

- Give participants time to express their own views, feelings, and conclusions. Ask: *"What happened in the game? Who has some view or feeling about the game that they would like to share?"*

Second, examine the causes of events and outcomes in the game by moving to a set of related questions about underlying assumptions, paradigm change, control, and the ethics of equity and inclusion:

- *"Did you assume at the beginning of the game that each person had to have his or her own loop? If so, why?"*
- *"Is it acceptable to develop a strategy that depends on some people dropping out?"*
- *"Did you take the time during the game to discuss longer-term strategy? If not, why not?"* The typical response is that players felt the facilitator was pushing them from one round to the next by announcing "Switch." If you hear this reply, ask the next question.

- *"How did those of you inside the circle feel about those who were not able to find a space? Who was responsible for outsiders' failure to find space?"*
- *"When the perception of diminishing resources occurred, how did people feel? Often there is a sense of giving up. What behavior does that transform into? Do we tend to see more innovation and creative problem solving or less?"*
- *"How did those of you outside the circle feel about those inside? Who was responsible for the fact that no one offered to help you find a place?"* Ask those who were removed from the game how they felt.
- *"To succeed in this game, you need to experience two paradigm changes, or strategy shifts. First, you have to recognize that each person does not need to have his or her own loop; people can share resources. Second, you must figure out that your entire foot does not need to be touching the ground. In this exercise, how did these shifts occur? Who first had these ideas? Was it someone who already had a space? Someone who was excluded? What other characteristics did the initiators of these ideas have? How did other group members respond to these shifts? Did they support them? Resist them? If the group supported these changes, what was it about the initiators that caused others to support their ideas?"*
- *"It was obvious early on that there would not be enough loops for everyone to find his or her own space using the policies with which you started the game. When the future limits became obvious, did group members change immediately or wait to innovate until they had no alternative? If they waited, why? What are the costs of dealing with limits only after they are pressing hard on the system? How could you change the system to make it anticipate limits and innovate in advance of absolute necessity?"*

Third, you can lead your participants through a discussion about lessons the game provides for dealing with climate change.

- *"How does your game behavior relate to climate change? Do the causes of the game results also exist in the real world? Generalize from these conclusions to predict the most likely sources of new ideas for controlling climate change?"*

The group processes you experienced in this exercise are quite common. In your script you may have talked about vanishing carbon sinks, habitable land, low-lying islands, arable land, water, and wildlife species. How else is climate change creating the dynamic of ever-diminishing resources? Are resources declining in the organizations that are dealing with the climate change issue? What might you learn from this exercise that you could use in your work on climate change to facilitate the spread of constructive new ideas?

You might comment, if you did indeed observe this, that in the end the game produces encouraging results that people can creatively find a way to share and cooperate, even with reduced resources.

17.

Squaring the Circle

Without a shared goal, cooperation is ineffective.

Quotes

We have it in our power to begin the world over again.

—Thomas Paine, philosopher

These civilizations and cultures that have governed our sense of the sacred and established our basic norms of reality and value . . . are terminating a major phase of their historical mission. The teaching and the energy they communicate are unequal to the task of guiding and inspiring the future. They cannot guide the great work that is before us. . . . Something new is happening. A new vision and a new energy are coming into being.

—Thomas Berry, ecotheologian

The ability to self-organize is the strongest form of resilience. A system that can evolve can survive almost any change, by changing itself.

—Donella Meadows, environmental leader

Climate Link

To avert the most serious effects of climate change, we are faced with the intense need to make significant changes in the fundamental structures on which our lives, political institutions, economies, and lifestyles depend. Climate change, however, is characterized by future surprises and unknowable risks. We must therefore build resilience and social learning into our own systems, so that society can self-organize and create a shared vision for working within, while fighting to change the drivers for, our fast-changing environment.

About This Game

During Squaring the Circle a team engages in a process that may feel a lot like real life—trying to develop a totally shared view of a problem and shared vision of the solution without any individual having a complete overview of the situation. In this exercise participants are literally in the dark.

In nature and in society, successful systems are those that possess a self-organizing capability—those that can act autonomously, view themselves in relation to their environment, and adapt accordingly. Squaring the Circle challenges a group to become its own self-organizing unit and find its own order through teamwork, shared visioning, visualization, and systemic thinking.

The purpose of this exercise is to explore experientially the meaning of social learning and to introduce the concept of self-organization, especially when a complete overview of the situation is not available. Deprived of sight and thus of a wide array of nonverbal communications (such as gestures and facial expressions), the group must adapt to its new environment, a challenge faced by all self-organizing groups. The game will also stimulate learning around some of the challenges of communication, the process of creating a shared vision, and joint problem solving.

To Run This Exercise

Number of People

This is a participation game. The minimum is eight people and the maximum is thirty. If you have more than thirty participants, you can run this game in multiple groups as long as you have sufficient ropes, a large enough space, and monitors to ensure the safety of each participant.

Time

Twenty to thirty minutes.

Space

Outdoors or in a room large enough for participants to form a loose circle and not be too close to stationary objects, walls, or furniture that could pose safety issues, as the group members walk slowly around blindfolded.

Equipment

One long rope, about 10 yards (9 m) or longer;
Blindfolds, if using.

Setup

Have the rope nearby and make sure you can uncoil it easily without having to unravel tangles and snarls. Ideally, it should already be uncoiled and laying on the floor.

Instructions and Script

Step 1: Have everyone line up shoulder to shoulder in a straight line, all facing in the same direction. Ask the participants to put their hands out in front of them, palms up. Place one end of the rope in the hands of a person at the end of the line and walk down the line having each person take hold of the rope with both hands. At the end of the line, turn around and walk back up to the original end, but this time just lay out the rope on the floor. Then tie the two ends of the rope together. Now all people are bunched on half of the loop.

Step 2: Tell participants the rules: *"Close your eyes (or put on your blindfold) for the rest of the task. I'll explain the task in a moment. The entire rope needs to be used. You may slide along the rope, but you cannot change positions with anyone else on the rope. When you personally think that the group has finished its task, raise your hand and I will ask for a vote. If a majority of the group thinks you are finished, I will ask you to stop and open your eyes. If only a minority join you in raising their hands, I will tell you all to remain with your eyes closed and keep going."*

Participants will typically ask at this point whether they are allowed to talk. A good reply is: *"Everything that is not forbidden is permitted."* If a participant doesn't want to close his eyes or accidentally opens them during the exercise, ask him to let go of the rope and step back silently. He will serve as an observer who can later help the group understand the strengths and weaknesses of their problem-solving approach. You can also ask one or two people to volunteer to act as observers prior to the start of the exercise.

Step 3: Say, *"Your goal is to rearrange your group into the shape of a square while everyone maintains his or her hold on the rope."*

Step 4: As the facilitator, you should see that no member of the group wanders into anything, such as a wall, a tree, or a hole. As the group attempts to solve the problem, you should remain silent. When

a participant raises a hand to signal that they think the process is complete, the facilitator asks the group to keep their eyes closed and vote on whether they are finished. If fewer than half the people think the task is finished, tell them to keep their eyes closed and continue working to achieve the goal. If the majority believes the task is accomplished, ask everyone to open their eyes. Have them place the rope on the ground, being careful to maintain the shape.

Step 5: Give the group a chance to look at the shape of the rope and then move to a comfortable place to sit and debrief. Leave the rope on the ground so that the group can refer to it during the debrief.

Debrief

Some groups create a perfect square, some a triangle, and others a shape that looks like an amoeba. Whatever the shape, you and the group can be certain that there is learning to be gained. If you had any observers, allow them to comment on what they saw. Then ask participants to describe their experience:

- *"How were features of this task similar to the challenge society faces in preventing and adapting to climate change?"*
- *"How easy was it to complete the task and solve the problem together (to square the circle)? What if the problem we were solving was climate change?"*
- *"What were some of the features of the process that helped the group accomplish this together? What features of the process hindered this?"*
- *"What was your strategy?"*
- *"Was the strategy effectively communicated?"*

Their strategies will vary. Some group members will figure out that they can make the process easier if they count off and try to align the group so that there are an equal number of participants on each side of the square (all sides are equal). Other groups will figure out that the process can be improved if the people who make up the corners are chosen. Very rarely, a small group will make a square out of the rope they control while simply ignoring the others. Although this latter strategy violates the rule that the entire rope needs to be used, it offers an interesting analogy for debriefing, so we do not normally intervene to prevent it.

Squaring the Circle provides a good opportunity to explore how the group may have learned over the duration of the exercise. Revisit what happened in the first few minutes of the activity: How does this compare to what was happening toward the end? How did the group improve?

To explore the concept of self-organizing groups, you might consider these questions:

- *"Did a leader emerge? Is it hard to lead when you don't have a 'solution'? What can leaders in this situation provide (for example, process leadership)?"*

- *"How did the leader or lack thereof affect the group dynamics?"*
- *"How did not being able to see affect the ability to communicate?"*
- *"How does learning occur in the climate change field? What examples can you identify?"*

Group administrators have often witnessed this somewhat ironic occurrence: after moving almost immediately and effortlessly into a good square, members of the group start to analyze and intellectualize the process, and their satisfactory solution deteriorates. Their final square becomes more misshapen than the group's original solution.

Link this game with the breadth of the change that needs to happen across society to make progress on averting the worst damages from climate change.

You may also wish to explore the analogy to communicating about a goal without necessarily being able to see one another. How are the dynamics similar for organizations, sectors, or communities working on climate change projects and communicating only virtually with one another?

Link this game to society's need, at many levels, to solve shared problems and define collective strategies around climate change. What reflections does the group have on how their process relates to the social change needed within the climate change context? What lessons can they draw?

18.

Thumb Wrestling

Life is not a zero-sum game.

Quotes

We must, indeed, all hang together or, most assuredly, we shall all hang separately.

—Benjamin Franklin, inventor and statesman

The only thing that will redeem mankind is cooperation.

—Bertrand Russell, philosopher

Focus on competition has always been a formula for mediocrity.

—Daniel Burrus, technology forecaster

Climate Link

Most members of the global community recognize that it is necessary to reduce the amount of greenhouse gases emitted into the atmosphere.

But most polluters assume that they can get someone else to make the short-term sacrifices associated with lower emissions. The rationalizations differ. Poor countries claim the problem was caused by the rich so the industrialized nations should make the biggest reductions. Populous nations ask for a quota to be defined per person; sparsely populated nations prefer that the quota be allocated to each country. Developing nations ask for the technologies they need to develop energy from non-carbon sources. Rich countries proclaim that they need more time to make big changes.

And so it goes. A fantasy has arisen that someone else can solve the problem, that cooperation and shared sacrifice will not be necessary, that a person will be better off if others can be persuaded to make all the necessary cuts. The ineffectiveness of recent climate change summits illustrates that each country posits that it is in a competitive situation with other nations and assumes it can be a free rider, benefiting from changes made by others without incurring major costs itself. This game creates an opportunity to examine assumptions about the potential of cooperation versus competition.

About This Game

It is one thing to talk about mental models and another to see them in action. Thumb Wrestling gently and humorously shows the consequences of the implicit assumption that life is a zero-sum game. People like this exercise because it is fun. But it can also lay the foundation for serious discussion.

To Run This Exercise

Number of People

This is a mass game that is played in pairs. If there is an odd number of people in one group—for example, an odd number of people in one row of the audience—one person might engage in a thumb-wrestling competition with two people at the same time or turn around and play with someone in the next row.

Time

Ten to twenty minutes.

Space

This game is typically played by audience members while they stay in their seats.

Equipment

None unless you use a marker and writing surface for the debrief.

Setup

The only setup required is the process of dividing the entire audience into pairs of opponents.

Instructions and Script

Step 1: Ask participants to find a partner, preferably by turning to the person next to them. If there is an uneven number, the leader may participate or one person can wrestle with two people simultaneously, using both the right and the left hand at the same time. It may go faster if you explicitly tell people how to find a partner or if you assign partners. But one way or another, you need to start by having each participant identify the person with whom they will play the game.

Step 2: Say, *"Now we are all going to engage in a simple competition called thumb wrestling. Over the next few minutes, your goal is to get for yourself as many points as you can. It is socially acceptable, during this exercise, for you to be completely selfish."*

Call up someone to help you demonstrate how to play the game. It is useful to pick someone who has been encouraged ahead of time to believe that it is best to employ an aggressive style of play. When your demonstration partner is up in front of the audience, interlock the fingers of your right hand with the fingers of his or her right hand.

Say, *"During the game each of you will have the goal of getting as many points for yourself as you can. To score each point, you must briefly pin, or press, the thumb of your opponent."*

Now engage in a few seconds of elaborate feinting and struggling with your demonstration partner to illustrate an aggressive process. Make sure that at some point you pin the other person's thumb between your thumb and the middle of your index finger. Pause and hold up your hands. Referring to the pinned thumb, *"That would give me one point. But since I want many points, I will immediately let loose and try to pin my opponent's thumb again."*

Step 3: After completing the demonstration, say, *"When I say, 'Go!' you will play the game for fifteen seconds. Each of you count your own points as you earn them. Be honest! Go!"*

Now let fifteen seconds elapse. The precise length is not crucial. You can either count to estimate the time or you can use a stopwatch. After the time has elapsed, say *"Stop!"*

Debrief

When the participants have stopped thumb wrestling, say, *"Now we will see how you did. Everyone who got three or more points for themselves, raise your hand."* Pause until people have had a chance to respond. Probably about half the audience will raise their hands. *"Thank you. Put down your hands. Now everyone who got six or more raise your hands."* Pause until people have had a chance to respond. *"Ten or more raise your hands."* Pause again. *"Fifteen or more. Twenty or more."* When only one or two couples are still raising their hands, ask them, *"How many points did you get?"* It will be a fairly large number, perhaps twenty to thirty. Repeat their answer loudly and with emphasis, so that the entire audience hears it. *"Please stand up and demonstrate your technique."* Almost certainly they will demonstrate a cooperative approach in which first one and then the other lowers his or her thumb so that the partner can quickly pin it. *"Thank you. Please sit down."*

Pause a moment while the participants sit down. Say, *"Obviously, the cooperative approach let both participants in the game get many more points for themselves than the competitive approach. Yet almost everyone here automatically assumed that they had to compete. They adopted a zero-sum attitude—'if you get more, I get less.' In fact the situation was win-win: either you both got many points, or you both got only a few. What similarities do you see between this exercise and climate change negotiations? How can you change the nature of climate change discussions so that nations are prompted to collaborate rather than compete?"*

If you can allocate an additional ten minutes to discussing this exercise, consider introducing the STUPID list to conclude the debrief. Say, *"You are an intelligent, socially responsible group of people. When a group like this one mainly adopts a poor strategy, as you did, there must be some underlying structural reasons. Let us see what they are. What were the factors that caused most of you to fall automatically into a competitive mode of behavior?"* Now pause and give them a minute or two to reflect on this. At this point, pick up a marker and stand by a large writing surface, a blackboard or flip chart. Usually someone will volunteer some explanation. Your job now is to prompt the group to come up with a variety of different factors while you plausibly rename them and write them down in a vertical list to give you the six phrases you need to spell out the acronym "STUPID." If your participants volunteer only a few factors, you can fill in the remainder. Typically the factors will not be offered in the required order, so leave blank lines when necessary to accommodate later entries. You must end up with the list below:

Small goals
Time pressures
Uncooperative partner
Poor vocabulary
Inadequate examples
Dysfunctional norms

Once you have written down the above list, look at the audience and point out that these factors not only apply to the thumb-wrestling exercise but also characterize most negotiations related to climate change.

"When we are operating under these conditions, we cannot expect positive results. Smart people can make bad decisions when they have small goals, time pressures, lack of cooperation, words poorly suited for discussing the problem, no examples of success, and social norms that are dysfunctional." Now draw a single long vertical loop around the first letter of each of the six factors and give everyone in the audience time to notice that these letters spell out "STUPID."

Point out that these factors mainly lie under our control, and we are free to change them. It then becomes useful to discuss how we could modify them in our efforts to deal with the climate.

19.

Triangles

**If you want big changes, look for
the high-leverage points.**

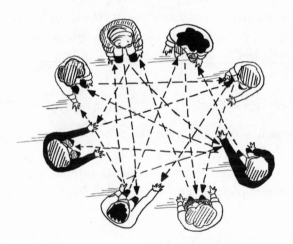

Quotes

There are high-leverage strategies that can help teams and individuals
deal with each challenge separately. But the greatest leverage comes
from understanding them as an ensemble of forces.

–Peter Senge, systems scientist

When we try to pick out anything by itself, we find it hitched to everything else in the Universe.

—John Muir, naturalist

It would be difficult to exaggerate the degree to which we are influenced by those we influence.

—Eric Hoffer, social philosopher

Climate Link

Climate change is being caused principally by direct and indirect consequences of the rising levels of greenhouse gases in the atmosphere. But any effort to modify behavior that leads to greenhouse gas emissions has many other repercussions. Most of the policies we enact eventually have very little impact on the GHG emissions we are trying to reduce. If we are to be effective in our efforts to change, we need to become skilled at identifying the few policy options among many that might actually lead to sustained change.

About This Game

Most people immediately grasp the concept of leverage points but have difficulty spotting them in an actual system. This exercise quickly illustrates the concept of leverage points through concrete changes made to the group's structure.

To Run This Exercise

Number of People

This is a participation game. A group of ten to forty people is best. If your group exceeds fifty, split it into groups with fewer than forty people in each.

Time

Twenty to thirty minutes.

Space

A square or rectangular area free of obstacles and at least 30 feet (9 m) long on each side, large enough for all participants to move around freely.

Equipment

None.

Setup

Ask the participants to stand in a single, large circle within the space you will be using for the exercise. Each person should be able to see all the others. You should stand inside the circle.

Instructions and Script

Step 1: Say, *"This exercise illustrates that some decisions have large impacts on the behavior of a system, while other decisions have no influence at all. I ask each of you to choose two other people around the circle who will be your references during this exercise. The first of your references should be someone who is wearing _____."* Here you should indicate a feature easily visible on one and only one person in the room, for example, blue eyeglasses, a yellow dress, a shirt that is hanging outside a belt, or red tennis shoes. Be careful to find some basis for selection that is not likely to embarrass the person who exhibits it. Items of clothing are generally acceptable. Physical features are best avoided. Psychological features, such as always being happy, should not be used since of course they are not obvious. Because participants typically will not know everyone else's name, you also cannot ask them to choose a reference on the basis of a unique name. The script below assumes you've chosen an item of clothing.

Say, *"You can't pick yourself as a reference. Thus, if you yourself are wearing [the feature], you should pick another person as your first reference. The second of your references can be anyone else you select, except that no one should choose themselves as a reference, nor should they choose a person who is wearing _____."* Here you name some characteristic that is exhibited by at least one other person in the exercise. If you have done this correctly, everyone in the group will have chosen the same specific person as their first reference (except that person, who cannot be his or her own reference). In the remainder of these instructions, we will call this person the universal reference.

And there will be one or more people in the room no one has picked as a reference. We will call everyone in this group a null reference.

Step 2: Once everyone has selected a second reference, say, *"Each time I say, 'Go!' your goal will be to move around the room slowly until you are equally distant from your two references. You can be very near or far from each of them; that makes no difference. But you should not stop moving until you are the same distance from each. When you are equidistant from your references, you should stop moving. But of course if one or both of your references move again after you've stopped, you, too, might need to move again in order to place yourself equidistant from the two of them.*

"If I say, 'All stop!' then everyone should halt immediately and remain standing in place. Any questions?"

To show what you mean, invite the universal reference and a second person to join you in the middle of the circle. Tell each of them to stand in specific locations, at least 5 feet (1.5 m) apart. Then pretend they are your references and illustrate how you would have to move to make yourself equidistant from them. Illustrate standing both close and equidistant and far away and equidistant. When you are equally distant from both of them, ask one of your two references to move several feet and then show how you would have to move to regain equidistance before you can stop. Again, ask if there are any questions.

Step 3: Return your two volunteers to their original positions, and move yourself out of the circle.

Say, *"In a minute we will do the first trial. I am going to ask everyone to get equidistant from their references. What will happen? Will the group keep moving forever, or will it come to a stop? How long will it take for the group to stop?"*

These questions are important. After each inquiry, pause and give your participants time to form their own answers. When they volunteer their replies, do not be judgmental; just thank them for their opinions. It is essential to get the participants to ponder and decide what they themselves think the behavior of the system will be. That will maximize their learning from their observations about what actually does occur.

Step 4: After you have given the participants a few minutes to speculate say, *"Trial number one. Go!"* Watch what happens. People will move out of their position in the circle and mill around the room for a few minutes before everyone slowly comes to a stop. Be patient. It may take longer to reach equilibrium, but the participants typically will come to rest. If they are still moving after three minutes and it does not seem the group is going to reach equilibrium in the near future, say, *"All stop. All of you have reached a satisfactory distance from your references."*

Now you can conduct two more trials that illustrate the concepts of high and low leverage, showing the extent to which a change in one part of the system can have impacts on a part ostensibly quite far away.

Step 5: Ask everyone to return to his or her original positions around the large circle. Now point to the person who is the universal reference. Announce that at some point during the next phase of the exercise you will use your hands to stop his or her movement. When you do that, everyone else should continue to follow the original rules, moving until they have achieved equidistance from their two references.

Ask what will happen. Give them time to reflect and respond. Do not evaluate the responses. When people are finished with their comments, you can say, *"Let's do an experiment and see."* Then say, *"Trial two. Go!"* Then quickly and gently stop the motion of the universal reference by laying your hand on his or her shoulder. The rest of the group typically will also stop very quickly. Return to the original circle and discuss what happened.

Step 6: Point to someone in the null reference group. Explain that in the third trial you will intervene by placing your hands on the shoulder of that person, causing him or her to stop moving. All others will continue to follow the rules—in other words, they will move until they are equidistant from their references. Ask the participants how your intervention will change the behavior of the system. Again, give them time to reflect and respond without evaluating their responses. Once the comments are finished, once more say, *"Let's do an experiment and see."*

Tell the group, *"Trial three. Go!"* Then quickly and gently stop the movement of the person from the null reference group you have previously indicated. Halting this person will not have any influence on the remainder of the group. It will still take some time for the rest of the group to come to a halt.

Return to the original circle. Ask what happened and why.

In the second of the three trials you used a high-leverage policy: you changed something that had influence on every other part of the system. In the third of the three trials you used a low-leverage policy: you changed something that had little influence on any other part of the system.

Debrief

Ask group members to take their seats and share their general impressions, feelings, and observations about the exercise. Many insights should already have emerged during the conversations after each experiment. Now it is time to summarize. Here are some possible questions to ask:

- *"In this exercise the indicator of impact is the length of time required for people to stop moving. What is the corresponding indicator of impact in the climate system?"*
- *"What are some of the low-leverage policies people are using in response to climate change? Why do they have so little influence?"*
- *"What are some of the high-leverage policies people could be using in response to climate change? Why would they be effective?"*
- *"What can we do to shift political activities more toward the high-leverage policies?"*
- *"What are the factors in society that behave as universal references? What are factors that none of us uses to determine our own behavior?"*

20.

Warped Juggle

Incremental changes produce improvements;
structural changes produce transformation.

Quotes

Incrementalism guarantees only one thing which is mediocrity.

—Faisal Khosa, physician

If you want something new, you have to stop doing something old.

—Peter Drucker, management consultant

He who doesn't risk never gets to drink champagne.

—Russian proverb

Climate Link

Attempts to solve the climate change crisis involve various actions that are perceived to have high impact, such as increasing vehicles' fuel efficiency, reducing chemical use in food production, replacing inefficient lightbulbs, and recycling waste materials. A few initiatives will have significant impact on the problem; most will not. Time and resource limits compel a search for the most effective policies. This search can be facilitated by refining the vocabulary used to evaluate alternatives. One approach is to differentiate between incremental and structural changes.

For example, incremental change is achieved by raising vehicle fuel efficiency; structural change would be achieved by moving more workers to houses near their place of employment. Improved gas mileage often permits people to afford larger cars, so there is finally little net reduction in emissions. Relocation can permit workers to sell their car and commute on foot or by bicycle That gives a permanent reduction in emissions. Incremental change in industry might come from altering packaging standards; structural change would come from shifting the composition of products so that products are fully recyclable.

About This Game

This exercise illustrates the difference between high and low leverage policies. It provides participants an opportunity to:

- Experience the difference between incremental and structural change;
- Become students of their own behavior as they observe their own habitual ways of forming assumptions;
- Use and examine the process for creating alternate solutions;
- Learn new words that will be helpful in evaluating alternative approaches to climate change;
- Point out to the participants their implicit assumptions about team learning and problem solving, so they can become more effective.

To Run This Exercise

Number of People

With small groups this can be conducted as a participation exercise. With larger groups it serves in demonstration mode. Running the game requires a minimum of six people and a maximum of twenty. This exercise works best with eight to twelve participants.

Time

Twenty to forty-five minutes.

Space

This exercise can be conducted indoors or outside, as long as there is space large enough for participants to stand shoulder to shoulder in a circle.

Equipment

Three tossable objects, for example, Koosh balls, stuffed animals, or rubber chickens. Avoid tennis balls, as they can be difficult to catch.

Setup

Have the three tossable objects on hand. If possible, show only one object at first, hiding the other two in your pocket or a bag.

Instructions and Script

Step 1: Gather the group into a circle, with you as a participating facilitator. Show one of the objects and tell the group that their first step is to set up a pattern for tossing the object around the group until everyone has had a chance to catch it and throw it on.

You may offer one or more members of the group the role of observer. You will eventually ask each observer to share their observations of the group's process.

Ask players to hold their hands out in front of them until they have received the ball, then to lower their hands after they have passed the ball on to another participant. That way, when someone is looking to throw the ball in order to establish the order the first time, he or she should consider throwing it only to someone whose hands are still stretched out. When all participants have received the object once, it should be returned to you.

Tell them they need only remember who threw the object to them and to whom they threw it. Begin by tossing the object to another member of the circle (but not to the person standing next to you). It is important to use a gentle, underhanded toss. This is not an exercise that should require expert catching skills. Slow the pace if necessary so everyone is comfortable with tossing and catching the object.

Step 2: The person receiving the object tosses it to someone else who has yet to touch it. When all members of the group have touched the object, it is tossed back to you. After the participants have thrown the object around the first time to establish the order, simulate the order once by having people point, one after the other in the same order that the object will go, to the person to whom they will throw the object. Allow the group to throw the first object using the established pattern until you are sure that they remember the sequence. Once that pattern is well established, you can stop and show the group the further two objects. Tell them that in the exercise you will start each of the three objects around the circle, one after the other.

Step 3: Ask the group: *"How long do you think it will take to toss all three objects in the sequence we have established?"*

Before coming to a consensus on the time, you should state that there are only two rules:

1. Everyone must touch each of the three objects once.
2. Each object must be touched once by each member of the group in the same sequence that the participants established in step 1.

When participants ask for clarification, it is important that you state there are only two rules (as outlined above). If participants want to know how they might bend the rules, simply repeat those two rules. Also, find out whether anyone has done this exercise before. If they have, ask them to participate but remain silent.

Step 4: With one of the participants acting as a timer, preferably using a digital watch, pass all three objects through the specified sequence. When all the objects have been returned to you, call *"Stop!"* and ask the timer how long it took. Whatever time you end up with, challenge the group to cut that time in half. You can spur groups on by saying their major competitor has done it in half the time they required. The exercise is complete when the participants feel they have done it in the fastest time possible.

Debrief

The exercise offers the group an opportunity to implement and assess two structural changes, each drastically reduces the time to attain the goal, and several incremental changes, which together have relatively little benefit.

For the first structural innovation group members perceive that they should alter their position around the circle, so that they stand next to the person who is their designated catcher. A shuffling then ensues until people are able simply to pass, rather than throw, the object from person to person. This change drastically cuts down on tossing time and minimizes drops and misses.

For the second structural change, participants realize that they can place the three objects on the floor and just touch each of them in the correct sequence. This change also markedly improves their performance. The remainder of their time and effort will be devoted to incremental changes, such as standing closer together and throwing the objects faster. Collectively these incremental actions provide relatively little benefit.

In Warped Juggle the constraint is very often the group's assumption that there are more rules than those stated by the facilitator. What is the limiting action? The limiting action here can be that participants persist in using the same approach without stopping to reflect on their assumptions, listen to other ideas, or consider other options.

The group experienced how immediate success can produce subtle constraints, particularly in the thinking of individuals and groups.

- This game is about setting targets. *"What were the different targets that were set for this game? If we consider targets related to climate change, leaders may choose to set very low targets (such as slowing the rate of growth of CO_2 emissions). However, as we saw in this game, if leaders encourage people to get only a little better, they will achieve that and be satisfied. But when a very challenging target is set, people more often seek revolutionary solutions. In what other ways do these observations regarding targets parallel the climate change discussion?"*

- *"What kinds of inherent pressures and constraints may accumulate in your organization or more broadly as a result of successes related to climate change?"* Possible constraints: Financial resources? Capacity to respond to inquiries? Number of staff?

- As a facilitator, you can also point out that the way in which we receive information affects the assumptions we make about that information. In this exercise the facilitator begins by tossing the object across the circle. Participants assume that they, too, have to toss the objects, even though nothing in the rules explicitly requires them to do so. The fastest times are actually achieved by not tossing the objects but placing them on the ground where group members can touch them in the correct sequence.

- Ask, *"How does the way information about climate change is conveyed affect the assumptions made about that information?"*

21.

Web of Life

To better understand systems, make the interconnections visible.

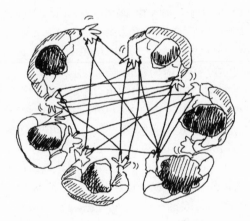

Quotes

All persons are caught in an inescapable network of mutuality, tied to a single garment of destiny. What affects one directly, affects all indirectly.

—Martin Luther King, Jr., civil rights leader

We need to stop thinking about these issues in isolation—each with its own champion, constituency and agenda—and deal with them in an integrated way, the way they actually occur on the ground.

—Glenn Prickett, Chief External Affairs Officer
with The Nature Conservancy[12]

When spider webs unite, they can tie up a lion.

—Ethiopian proverb

Climate Link

It is easy to think that our actions exist in isolation. Grow more food if you need to feed a growing population, right? But clearing land for agriculture will reduce the ability of forestlands and wetlands to store carbon. That will eventually produce higher GHG concentrations, with negative effects on future food production. Climate change is characterized by high behavioral and dynamic complexity, yet solutions offered are often fragmented, focusing on a part of the puzzle rather than the intricate web of interrelationships that make up the whole of the climate change system.[13]

Given the weblike nature of climate change, where do you start? Do you develop climate change policy and regulations? Develop communications strategies to motivate the general public or specific groups to change behaviors? Create a set of indicators to help communities monitor and make community-wide changes? Work with businesses to develop pro-climate investment strategies?

You can use this exercise to physically trace the interconnections and dynamics of many social and economic policies. You can also use the Web of Life exercise to explore the physical science of climate change.

About This Game

The Web of Life game allows a group to observe how the parts of a system of interest are interconnected. As they go through this exercise, participants see that the system at hand, whether the physical climate system or the behavioral system related to a specific climate change intervention or policy, is typically not made up of straight lines of causality but patterns of connection and interaction that resemble loops, webs, and networks. In most situations we can't see these interconnections; we have to imagine them. Web of Life gives participants an opportunity to appreciate and make visible the often complex patterns of interconnection that compose some of the more perplexing challenges associated with global warming.

In addition, this exercise can help teams see and better understand the interdependencies and the connections that exist among themselves and within the larger system (such as their organization and their community).[14]

To Run This Exercise

Number of People

This is a demonstration game. A group of eight people works well.

Time

Fifteen to thirty minutes, depending on the number of people.

Space

Enough space so your group can stand in a circle nearly shoulder to shoulder.

Equipment

One large ball of colorful yarn or string (make sure it will unravel easily);
Flip chart or other writing surface;
Packet of sticky Post-it notes (or larger pieces of paper and tape, for a bigger group).

Setup

Organize the group into a circle. Give someone in the group the ball of yarn.

Instructions and Script

Step 1: While you and your group are standing in a circle, ask group members to pick a climate issue they wish to discuss. For example, you could use this exercise with local stakeholders who are challenged to understand the necessary agricultural adaptations in a specific region as a result of changing climate conditions.

Step 2: After you have picked your issue, begin to brainstorm some of the variables related to that system. List the variables on the flip chart as people identify them. Then write the variables on Post-it notes or make bigger signs if a larger group is involved. Give one variable to each participant to put on like a name tag. Using agricultural adaptations to climate change, as an example, the group might brainstorm the following variables:

- Global temperatures
- Weather extremes
- Water shortages
- Food supply
- Amount of heat in the atmosphere
- Crop and livestock yields
- Amount of exports

Step 3: Start by having the person holding the yarn name his or her variable. For instance, the person may pick "global temperatures."

Step 4: Someone else in the circle then names his or her own variable and tells how it is related to that of the person holding the yarn. For instance, the participant states, "If global temperature rises, weather extremes become larger." The second player then accepts the ball of yarn from the first.

Now another person explains how his or her variable is related to that of the person with the yarn. For example, the player takes the ball of yarn, saying, "If weather extremes become larger, crop and livestock yields become smaller."

Step 5: The group continues identifying as many connections as possible while the web grows in complexity. Once the group is sufficiently entwined, ask, *"Have you captured most of the important*

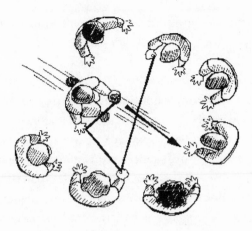

relationships?" Then have the participants place the web, intact, on the floor where they are standing and return to their seats.

Debrief

Listen carefully to the group's comments as the web becomes more intertwined and complicated. Remember or write down a few of their comments. After they have finished, ask the group for their reactions.

- *"Which variables have the most connections? What does that tell you?"*
- *"What is the length of the time horizon we must consider? Looking at the web, where do you see significant time delays between cause and effect, between actions and their consequences? How long are those delays?"*
- *"How might this dense network of interconnections influence groups' abilities to take action?"*
- *"Where do you see significant nonlinearities between actions and consequences, instances when small deviations from a norm produce no response but a slightly larger deviation produces a dramatic change?"*
- *"How might this exercise be used to reveal and communicate to the appropriate research or decision-making groups their*

existing knowledge about the interconnections and dynamics inherent in agricultural adaptations?"

Ask, *"Is climate change really the problem?"* For example, how is climate change a symptom of another problem (for example, economic growth)? Work with the group to brainstorm eight or more variables related to economic growth: number of people, amount of resources consumed, amount of waste generated, amount of greenhouse gases, land use, habitat loss, and so on. Now using the ball of yarn, explore how the interaction of these factors may produce symptoms such as climate change: *"Looking at the web you have created, where do you see leverage for fundamental change?"*

22.

1-2-3-Go!

Actions speak louder than words.

"1-2-3..."

Quotes

To do good is noble. To tell others to do good is even nobler and much less trouble.

—Mark Twain, humorist

Mixed messages are highly damaging to public understanding, trust and sense of personal capacity to act.

—Ian Christie, Green Alliance

A man is judged by his deeds, not by his words.

—Russian proverb

Climate Link

Climate activists realize that the worst extremes of climate change will only be avoided by widespread changes in behavior—in consumption, politics, travel, recreation, production, energy use, and so on. The necessary changes are not occurring now, and there is no chance they will, unless advocates and activists become more effective in their exhortations. Effective climate action requires a much better understanding of the relation between words and deeds.

About This Game

This exercise demonstrates that leading by example provides more power to your words and messages. It supports the popular saying: "Actions speak louder than words."

This exercise is relatively brief, so it works effectively as an introduction to a workshop or as a wrap-up activity. It makes the point that whatever we learn and whatever we promise to do, our constituencies, our organizations, and our social networks will be more influenced by what we do than by what we tell them to do.

As an example of how this exercise is used, Chirapol Sintunawa of Mahidol University in Thailand often tells participants in the closing discussion of his sustainability workshops that he does not want them to go home and *tell* anyone about what they learned during the day. He then introduces 1-2-3-Go! At the end of the exercise he asks them to demonstrate their learning through their actions, not exhortations. He convinces them that will have much more impact in the long term.

The exercise 1-2-3-Go! can be used as a wrap-up exercise at the end of your presentation. If used this way, it is important to do it quickly and use it lightly. You do not want the participants to consider it a trick you played on their group. For this reason, you may play it twice, giving people ample warning the second time and laughing along with the group if someone forgets and again claps at the wrong point. With caution and a light touch on the part of the facilitator, this simple exercise can help participants focus more attentively on the congruence of their words and actions.

A special advantage of this exercise is that it raises the impact of the game if a few people have seen it before and do it correctly.

To Run This Exercise

Number of People

This is a mass game. The minimum is two people, and there is no maximum, as any number can play.

Time

Three to ten minutes.

Space

Enough space so everyone can see the facilitator. Typically this game is conducted with all the audience members in their seats.

Equipment

None.

Setup

None.

Instructions and Script

Step 1: Ask everyone to be sure they can see you.

Step 2: While you demonstrate holding out your hands as if you were going to clap, ask participants to hold their arms out in front of them. Say, *"Now I am going to count slowly to three and then say 'Go!' When I say 'Go!' everyone should clap their hands together once* *and in the same instant."* Repeat these instructions to make sure that everyone hears and is paying attention to you. You can add, *"Our goal here is for everyone to clap at the same time, so that it sounds like one giant pair of hands clapping loudly. After all, we are all doing the same thing, and doing it together will amplify our individual impact."* Repeat the instructions at this point, and say, *"Now I am going to count to three, and say "Go."*

Step 3: Slowly count, *"One, two, three,"* then clap your hands together loudly, pause for one second, then say, *"Go!"*

When you clap your hands together, almost everyone will clap their hands together, not waiting until you say *"Go,"* as they were instructed. Pause a moment, and let everyone realize what happened. At this point have your debriefing discussion as below.

Debrief

For 1-2-3-Go! the debrief should be light and quick. Different points can be made, and it is interesting to ask the group at the end of the first trial what lessons they can take away from the exercise. After their points, you can make the following observation: *"This exercise points out the importance of nonverbal communication in determining what happens when you are trying to make impact or catalyze change. People will not only attend to what you say but will look at what you do. If you want your words to have the most impact, it is crucially important that your actions be consistent with what you are telling people."*

Here is a question to explore:

- *"What are the ways your behavior might send confusing signals to your constituencies, partners, or community?"*

This question does not necessarily need an answer. However, if your wrap-up discussion is focused on personal action in people's climate change work, you might get more thoughtful and personalized answers.

It can be a bit depressing to end with a game where almost everyone makes a mistake. So repeat the instructions and hope everyone will do it correctly. That shows that people have the capacity to learn from their mistakes and ends the game on a high note.

Acknowledgments

The authors thank the German Ministry for International Cooperation (GIZ) GmbH for their support, including the funding that permitted us to write the first version of this book. We are grateful for their help in testing the games and for their many valuable suggestions on relating our simple games to essential ideas about climate change.

Notes

1. Bryner, Andy and Markova, Dawna, *An Unused Intelligence: Physical Thinking for 21st Century Leadership* (Berkeley, CA: Conari Press, 1996).
2. The Arms Crossed exercise is very different from the version that first appeared in 1995 in *The Systems Thinking Playbook*. Over the years, the ways it is introduced, conducted, and debriefed have all evolved. ·
3. For more on public complacency about climate change, see John Sterman and Linda Booth Sweeney, "Understanding Public Complacency about Climate Change: Adults' Mental Models of Climate Change Violate Conservation of Matter," *Climatic Change* 80, 3–4 (2007): 213–238.
4. This exercise was adapted from Rob Quaden, Alan Ticotsky, and Debra Lyneis's "In and Out Game"; see http://static.clexchange.org/ftp/documents/x-curricular/CC2010-11Shape1InAndOutSF.pdf.
5. This exercise does not appear in the original *Playbook*. Dennis Meadows invented it as his contribution to a 2009 panel in Hungary on the relation of ecosystems to climate change.
6. Donella Meadows, *Thinking in Systems* (White River Junction, VT: Chelsea Green, 2008).
7. The period during which scientists were monitoring low ozone readings and yet not "seeing" them is described well in Paul Brodeur, "Annals of Chemistry: In the Face of Doubt," *New Yorker,* June 9, 1986, 71.
8. A more elaborate version of the Harvest game, FishBanks, is a computer-assisted, role-playing game for groups up to fifty. It takes two hours to play and is rich in learning. For a set of physical materials required to play FishBanks, contact the International System Dynamics Society, http://www.systemdynamics.org/products/fish-bank/. For the Internet version of FishBanks,

see https://mitsloan.mit.edu/LearningEdge/simulations/fishbanks
/Pages/fish-banks.aspx.

9. For an excellent discussion of the archetype of the tragedy of
   the commons, see Daniel H. Kim, *Systems Archetypes II: Using
   Systems Archetypes to Take Effective Action* (Acton, MA:
   Pegasus Communications, 1994).

10. The authors adapted the Living Loops game from an exercise
    created by John Shibley.

11. For more examples, see: "Feedback Loops in Global Climate
    Change Point to a Very Hot 21st Century, " available at http://
    www2.lbl.gov/Science-Articles/Archive/ESD-feedback-loops
    .html, and "Feedback Loops: The Potential to Amplify Global
    Warming Beyond Current Predictions," available at http://www
    .andweb.demon.co.uk/environment/globalwarmingfeedback.html.

12. Thomas L. Friedman, "Connecting Nature's Dots," *New York
    Times*, August 22, 2009.

13. In situations of high dynamic complexity, cause and effect are
    distant in time and space, and causes of problems are not easily rec-
    ognized through firsthand experience. Behavioral complexity is the
    extent to which there is diversity in the mental models, aspirations,
    and values of decision makers related to a particular challenge.

14. The original Web of Life exercise was inspired by Outward Bound.

About the Authors

Dennis Meadows is emeritus professor of systems policy and social science research at the University of New Hampshire, where he was also director of the Institute for Policy and Social Science Research. In 2009 he received the Japan Prize for his contributions to world peace and sustainable development. He has authored ten books and numerous educational games, which have been translated into more than thirty languages for use around the world. He earned his PhD in management from MIT, where he previously served on the faculty, and he has received four honorary doctorates for his contributions to environmental education.

Linda Booth Sweeney, EdD, is an educator, researcher, and writer dedicated to helping people of all ages integrate an understanding of complex, living systems into learning, decision making, and design. She has worked with Outward Bound, MIT's Sloan School of Management, and Schlumberger Excellence in Educational Development (SEED). She is the coauthor of *The Systems Thinking Playbook* and the author of *When a Butterfly Sneezes: A Guide for Helping Kids Explore Interconnections in Our World through Favorite Stories, and Connected Wisdom: Living Stories about Living Systems;* as well as articles for numerous academic journals and newsletters. She lives outside Boston, Massachusetts. For more information, see her blog, *Talking about Systems* (www.lindaboothsweeney.net/blog).

Gillian Martin Mehers is a learning and capacity development practitioner who has worked within the global sustainability community for over twenty years. She is a founder of Bright Green Learning, a Geneva-based professional collaborative that facilitates group learning about issues related to sustainability. She was previously the head of learning and leadership at the International Union for Conservation of Nature (IUCN) and prior to that the director of capacity development for an international organization based in London, Leadership for Environment and Development (LEAD) International. Mehers's expertise is in creating experiential learning environments, interactive learning designs, and process facilitations for diverse stakeholders who wish to improve their communication and learning. With a particular passion for working interculturally, she has been a facilitator and trainer in over fifty countries, from Armenia to Zambia. For more information, see her blog: *You Learn Something New Every Day* (www.welearnsomething.org).

green press
INITIATIVE

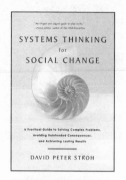

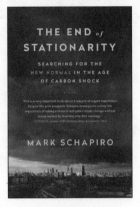
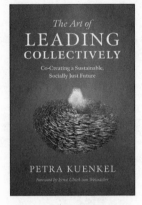